Andreas & Dagmar Hensing

MAKING A MORGAN

Also from Veloce ...

Enthusiast's Restoration Manual Series

Citroën 2CV Restore (Porter)
Classic Car Bodywork, How to Restore (Thaddeus)
Classic British Car Electrical Systems (Astley)
Classic Car Electrics (Thaddeus)
Classic Cars, How to Paint (Thaddeus)
How to Restore & Improve Classic Car Suspension, Steering & Wheels (Parish – translator)
Jaguar E-type (Crespin)
Reliant Regal, How to Restore (Payne)
Triumph TR2, 3, 3A, 4 & 4A, How to Restore (Williams)
Triumph TR5/250 & 6, How to Restore (Williams)
Triumph TR7/8, How to Restore (Williams)
Ultimate Mini Restoration Manual, The (Ayre & Webber)

Those Were The Days ... Series

Austerity Motoring (Bobbitt)
Austins, The last real (Peck)
Brighton National Speed Trials (Gardiner)
British Woodies (Peck)
Don Hayter's MGB Story – The birth of the MGB in MG's Abingdon Design & Development Office (Hayter)
MG's Abingdon Factory (Moylan)
Three Wheelers (Bobbitt)

Great Cars

Austin-Healey – A celebration of the fabulous 'Big' Healey (Piggott)
Jaguar E-type (Thorley)
Jaguar Mark 1 & 2 (Thorley)
Triumph TR – TR2 to 6: The last of the traditional sports cars (Piggott)

General

1½-litre GP Racing 1961-1965 (Whitelock)
AC Two-litre Saloons & Buckland Sportscars (Archibald)
Alfa Romeo 155/156/147 Competition Touring Cars (Collins)
Alfa Romeo Giulia Coupé GT & GTA (Tipler)
Alfa Romeo Montreal – *The dream car that came true (Taylor)
Alfa Romeo Montreal – The Essential Companion (Classic Reprint of 500 copies) (Taylor)
Alfa Tipo 33 (McDonough & Collins)
Alpine & Renault – The Development of the Revolutionary Turbo F1 Car 1968 to 1979 (Smith)
Alpine & Renault – The Sports Prototypes 1963 to 1969 (Smith)
Alpine & Renault – The Sports Prototypes 1973 to 1978 (Smith)
An Austin Anthology (Stringer)
Anatomy of the Classic Mini (Huthert & Ely)
Anatomy of the Works Minis (Moylan)
Armstrong-Siddeley (Smith)
Art Deco and British Car Design (Down)
Austin Cars 1948 to 1990 – a pictorial history (Rowe)
Autodrome (Collins & Ireland)
Automotive Mascots (Kay & Springate)
Bentley Continental, Corniche and Azure (Bennett)
Bentley MkVI, Rolls-Royce Silver Wraith, Dawn & Cloud/Bentley R & S-Series (Nutland)
BMC Competitions Department Secrets (Turner, Chambers & Browning)
BMW 5-Series (Cranswick)
BMW Z-Cars (Taylor)
BMW Classic 5 Series 1972 to 2003 (Cranswick)
BMW – The Power of M (Vivian)
British Cars, The Complete Catalogue of, 1895-1975 (Culshaw & Horrobin)
BRM – A Mechanic's Tale (Salmon)
BRM V16 (Ludvigsen)
Bugatti – The 8-cylinder Touring Cars 1920-34 (Price & Arbey)
Bugatti Type 40 (Price)
Bugatti 46/50 Updated Edition (Price & Arbey)
Bugatti T44 & T49 (Price & Arbey)
Bugatti 57 2nd Edition (Price)
Bugatti Type 57 Grand Prix – A Celebration (Tomlinson)
Carrera Panamericana, La (Tipler)
Citroën DS (Bobbitt)
Cobra – The Real Thing! (Legate)
Competition Car Aerodynamics 3rd Edition (McBeath)
Competition Car Composites A Practical Handbook (Revised 2nd Edition) (McBeath)
Cortina – Ford's Bestseller (Robson)
Cosworth – The Search for Power (6th edition) (Robson)
Coventry Climax Racing Engines (Hammill)
Daily Mirror 1970 World Cup Rally 40, The (Robson)
Daimler SP250 New Edition (Long)
Datsun Fairlady Roadster to 280ZX – The Z-Car Story (Long)
Dino – The V6 Ferrari (Long)
Ferrari 288 GTO, The Book of the (Sackey)
Ferrari 333 SP (O'Neil)
Fiat & Abarth 124 Spider & Coupé (Tipler)
Fiat & Abarth 500 & 600 – 2nd Edition (Bobbitt)
Fiats, Great Small (Ward)
Ford Model Y (Roberts)
Ford Small Block V8 Racing Engines 1962-1970 – The Essential Source Book (Hammill)
Grand Prix Ferrari – The Years of Enzo Ferrari's Power, 1948-1980 (Pritchard)
GT – The World's Best GT Cars 1953-73 (Dawson)
Hillclimbing & Sprinting – The Essential Manual (Short & Wilkinson)
Immortal Austin Seven (Morgan)
Inside the Rolls-Royce & Bentley Styling Department – 1971 to 2001 (Hull)
Intermeccanica – The Story of the Prancing Bull (McCredie & Reisner)
Jaguar from the shop floor (Martin)
Jaguar E-type Factory and Private Competition Cars (Griffiths)
Jaguar, The Rise of (Price)
Jaguar XJ 220 – The Inside Story (Moreton)
Jaguar XJ-S, The Book of the (Long)
The Jowett Jupiter – The car that leaped to fame (Nankivell)
Karmann-Ghia Coupé & Convertible (Bobbitt)
Lamborghini Miura Bible, The (Sackey)
Lamborghini Murciélago, The book of the (Pathmanathan)
Lamborghini Urraco, The Book of the (Landsem)
Lancia 037 (Collins)
Lancia Delta HF Integrale (Blaettel & Wagner)
Lancia Delta Integrale (Collins)
Land Rover Emergency Vehicles (Taylor)
Land Rover Series III Reborn (Porter)
Land Rover, The Half-ton Military (Cook)
Land Rovers in British Military Service – coil sprung models 1970 to 2007 (Taylor)
Lea-Francis Story, The (Price)
Le Mans Panoramic (Ireland)
Lexus Story, The (Long)
Lola – The Illustrated History (1957-1977) (Starkey)
Lola – All the Sports Racing & Single-seater Racing Cars 1978-1997 (Starkey)
Lola T70 – The Racing History & Individual Chassis Record – 4th Edition (Starkey)
Lotus 18 Colin Chapman's U-turn (Whitelock)
Lotus 49 (Oliver)
Maserati 250F In Focus (Pritchard)
Mazda MX-5/Miata 1.6 Enthusiast's Workshop Manual (Grainger & Shoemark)
Mazda MX-5/Miata 1.8 Enthusiast's Workshop Manual (Grainger & Shoemark)
Mazda MX-5 Miata, The book of the – The 'Mk1' NA-series 1988 to 1997 (Long)
Mazda MX-5 Miata, The book of the – The 'Mk2' NB-series 1997 to 2004 (Long)
Mazda MX-5 Miata Roadster (Long)
Mazda Rotary-engined Cars (Cranswick)
Maximum Mini (Booij)
Meet the English (Bowie)
Mercedes-Benz SL – R230 series 2001 to 2011 (Long)
Mercedes-Benz SL – W113-series 1963-1971 (Long)
Mercedes-Benz SL & SLC – 107-series 1971-1989 (Long)
Mercedes-Benz SLK – R170 series 1996-2004 (Long)
Mercedes-Benz SLK – R171 series 2004-2011 (Long)
Mercedes-Benz W123-series – All models 1976 to 1986 (Long)
Mercedes G-Wagen (Long)
MG, Made in Abingdon (Frampton)
MGA (Price Williams)
MGB & MGB GT– Expert Guide (Auto-doc Series) (Williams)
Mini Cooper – The Real Thing! (Tipler)
Mini Minor to Asia Minor (West)
Morgan Maverick (Lawrence)
Morgan 3 Wheeler – back to the future!, The (Dron)
Morris Minor, 70 Years on the Road (Newell)
Motor Racing – Reflections of a Lost Era (Carter)
Motor Racing – The Pursuit of Victory 1930-1962 (Carter)
Motor Racing – The Pursuit of Victory 1963-1972 (Wyatt/Sears)
Motor Racing Heroes – The Stories of 100 Greats (Newman)
Motorsport In colour, 1950s (Wainwright)
Porsche 356 (2nd Edition) (Long)
Porsche 908 (Födisch, Neßhöver, Roßbach, Schwarz & Roßbach)
Porsche 911 Carrera – The Last of the Evolution (Corlett)
Porsche 911R, RS & RSR, 4th Edition (Starkey)
Porsche 911, The Book of the (Long)
Porsche 911 – The Definitive History 2004-2012 (Long)
Porsche – The Racing 914s (Smith)
Porsche 911SC 'Super Carrera' – The Essential Companion (Streather)
Porsche 914 & 914-6: The Definitive History of the Road & Competition Cars (Long)
Porsche 924 (Long)
The Porsche 924 Carreras – evolution to excellence (Smith)
Porsche 928 (Long)
Porsche 930 to 935: The Turbo Porsches (Starkey)
Porsche 944 (Long)
Porsche 993 'King Of Porsche' – The Essential Companion (Streather)
Porsche 996 'Supreme Porsche' – The Essential Companion (Streather)
Porsche 997 2004-2012 – Porsche Excellence (Streather)
Porsche Boxster – The 986 series 1996-2004 (Long)
Porsche Boxster & Cayman – The 987 series (2004-2013) (Long)
Porsche Racing Cars – 1953 to 1975 (Long)
Porsche Racing Cars – 1976 to 2005 (Long)
Porsche – The Rally Story (Meredith)
Porsche: Three Generations of Genius (Meredith)
Powered by Porsche (Smith)
RAC Rally Action! (Gardiner)
Racing Colours – Motor Racing Compositions 1908-2009 (Newman)
Rallye Sport Fords: The Inside Story (Moreton)
Rolls-Royce Silver Shadow/ Bentley T Series Corniche & Camargue – Revised & Enlarged Edition (Bobbitt)
Rolls-Royce Silver Spirit, Silver Spur & Bentley Mulsanne 2nd Edition (Bobbitt)
Rootes Cars of the 50s, 60s & 70s – Hillman, Humber, Singer, Sunbeam & Talbot (Rowe)
Rover P4 (Bobbitt)
RX-7 – Mazda's Rotary Engine Sportscar (Updated & Revised New Edition) (Long)
SM – Citroën's Maserati-engined Supercar (Long & Claverol)
Standard Motor Company, The Book of the (Robson)
Toleman Story, The (Hilton)
Triumph & Standard Cars 1945 to 1984 (Warrington)
Triumph TR6 (Kimberley)
Two Summers – The Mercedes-Benz W196R Racing Car (Ackerson)
TWR Story, The – Group A (Hughes & Scott)
Unraced (Collins)
Volkswagens of the World (Glen)
VW Beetle Cabriolet – The full story of the convertible Beetle (Bobbitt)
VW Golf: Five Generations of Fun (Copping & Cservenka)
You & Your Jaguar XK8/XKR – Buying, Enjoying, Maintaining, Modifying – New Edition (Thorley)
Wolseley Cars 1948 to 1975 (Rowe)
Works Minis, The Last (Purves & Brenchley)
Works Rally Mechanic (Moylan)

www.veloce.co.uk

First published in 2015 by Heel Verlag under the title Morgan – Das Making-of. Published in English in 2015 and 2018 by Veloce Publishing Limited, Veloce House, Parkway Farm Business Park, Middle Farm Way, Poundbury, Dorchester DT1 3AR, England. Tel +44 (0)1305 260068 / Fax 01305 250479 / e-mail info@veloce.co.uk / web www.veloce.co.uk or www.velocebooks.com.
ISBN: 978-1-787113-69-5 UPC: 6-36847-01369-1.
© 2015 Heel Verlag. © 2015 and 2018 Andreas and Dagmar Hensing and Veloce Publishing. All rights reserved. With the exception of quoting brief passages for the purpose of review, no part of this publication may be recorded, reproduced or transmitted by any means, including photocopying, without the written permission of Veloce Publishing Ltd. Throughout this book logos, model names and designations, etc, have been used for the purposes of identification, illustration and decoration. Such names are the property of the trademark holder as this is not an official publication. Readers with ideas for automotive books, or books on other transport or related hobby subjects, are invited to write to the editorial director of Veloce Publishing at the above address. British Library Cataloguing in Publication Data – A catalogue record for this book is available from the British Library. Typesetting, design and page make-up all by Veloce Publishing Ltd on Apple Mac. Printed and bound by CPI Group (UK) Ltd, Croydon, CR0 4YY.

Andreas & Dagmar Hensing

MAKING A
MORGAN

17 days of craftsmanship:
step-by-step from specification sheet to finished car

VELOCE PUBLISHING
THE PUBLISHER OF FINE AUTOMOTIVE BOOKS

CONTENTS

PROLOGUE 6

THE MORGAN STORY 8

MAKING A MORGAN

DAY 1	Where it all begins: chassis shop	46
	Specifications	47
	Our story so far ... Conversation with Nick Baker on what happens after an order is received	48
	Steve Morris – the chieftain	50
	Brett Burbeck – the perfectionist	50
	Chassis Shop	51
	Machine Shop	52
	Chassis Shop II	56
	Break time	59
	Independent front suspension	61
DAY 2	The chassis is put on wheels	63
	Wood Shop	68
	Vince Wanklin – bedrock of the wood shop	69
DAY 3	'The elbow' – or: small cause, big effect	71
	A digression: From ordinary plywood to exotic wooden dashboard	75
DAY 4	The wooden frame is finished	78
	Just a word with Graham Chapman on 'quality'	81
	Sheet Metal Shop (aka the Tin Shop)	82
DAY 5	Sheet Metal Shop – second day	84
	Richard Harris – the 'Tin God'	85
	The press brake	86
	The first work week – our story so far	89
DAY 6	The scuttle panel	90
	A digression: production of the 3 Wheeler	94
	Assembly Shop	96
	Jason Hallett – the golfer	97
	A visit to Superform Aluminium in Worcester	99
DAY 7	Looking like a car now	102
	Marcin Olszewski – the Bonneteer	105
	Philip Jones – the Electrician	107
DAY 8	Connection works	109
	Mick Bishop – the athlete	113
	Kevin Bunn – the career man	114
	Paint Shop	114
	The colours of Morgan – an agony of choices	116
DAY 9	Paint prep work	117
	The Cole family – the 'real Morgans'	120
DAY 10	Paint Shop	121
	The second work week – our story so far	123
DAY 11	Third day in the Paint Shop	124
DAY 12	Trim Shop	128
	Leather for the Morgan	131
DAY 13	Trim Shop II	132
DAY 14	The soft top	136
	Ben Jones – the racer	137
	Tonneau cover	138
	Electrics and small parts	140
DAY 15	Finishing touches	145
	Test drive	146
DAY 16	Underseal	150
DAY 17	PDI	152

EPILOGUE 156

ACKNOWLEDGEMENTS 158

PREFACE

The information conveyed within this book details those qualities which position the Morgan Motor Company so prominently within the global automotive landscape.

The company is firmly established as a manufacturer synonymous with pure excellence, delivering a personalised service over the entire product lifecycle. This interaction with the brand is both personal and rich, providing customers with their own unique story.

The roots of this enjoyment stem from an illustrious history coupled with the way in which these vehicles are hand crafted, but most notably from the passion and commitment to this infamous iconic brand that is apparent in both the workforce and customers alike.

The ethos at Morgan remains unchanged; to provide an unrivalled motoring experience for their customers which starts from the moment they first think of becoming a Morgan owner, right through to taking delivery of their very own bespoke hand built luxury sports car.

This book encapsulates that experience and provides a showcase for the methods of manufacturing a car using traditional skills, highlighting how these techniques are harmoniously married with some of the latest technology to ensure we remain relevant to today's manufacturing environment whilst keeping the romance of traditional coach building alive.

Steve Morris
Managing Director

PROLOGUE

When I presented our idea to write a book on the making of a Morgan – from production planning through pre-delivery inspection – to Steve Morris, Managing Director of the Morgan factory at the 2014 Geneva Auto Show, it took him less than five minutes to grant his approval.

After a year of preparation, our project got underway in May 2015. We wanted to document the genesis of a traditional model, either a 4/4 or a Plus 4. Both are representative of the marque's roots, for its renunciation of all excess – with 'excess' defined as anything that is not absolutely necessary to the pure experience of driving. These 'excesses' include assistance systems such as power steering, or even electronically-controlled crutches such as ESP or ABS. Airbags, too, are of no use to any purist – at least, of no use in moving a vehicle from Point A to Point B. Strictly speaking, that would apply as well to safety belts and headrests, but those at least are required by law.

And so the Morgan – once production of the Land Rover Defender finally comes to an end in 2015 – remains as the last of a breed. The very last truly original automobile, or, to borrow the subtitle of our first Morgan book, "the last of the real sports cars."

The VW Beetle and the Land Rover remained in production over a longer span – the Beetle from 1938 to 2003, or 65 years (in Germany, only until 1978). The Land Rover was born in 1948, but was built by a succession of owners: Rover, British Leyland, Rover Group again, BMW, Ford, and now Tata.

Not so Morgan. The 4/4 and the Plus 4 have remained 'The Morgan' typified, in classic style since 1954, or, in terms of basic concept, since 1936. As such, these models leave in their dust any other automotive claimants to the title of longest production run. Moreover, the end of Morgan production is nowhere in sight. The 4/4 and Plus 4 are the two models that, then as now, represent the lion's share of the small specialist marque's production.

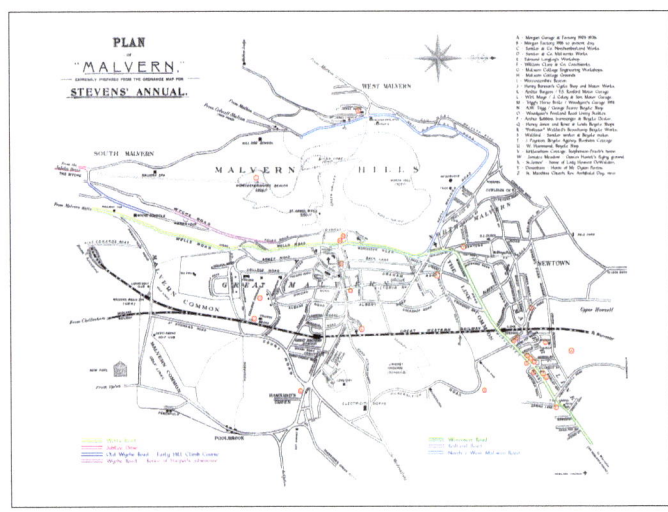

Malvern Link, c.1907.

Great Malvern at the beginning of the 20th Century.

Malvern Link, 2015.

From the very beginning, Morgans have been built piece by piece, purely by hand craftsmanship, in Malvern Link – a small town about 30 miles south-west of Birmingham – in a complex of brick buildings that has stood for more than 100 years. Over the decades, these have been supplemented by only a couple of assembly buildings.

Morgans are assembled by workmen – even today, there are very few women working in production – workmen who are still masters of traditional craftsmanship. These include

Church Street, Great Malvern, c.1860.

Church Street, Great Malvern, 2015.

The factory grounds, 2015.

carpenters, coachbuilders, saddlers and painters, and they are proud of their company. While accompanying 'our' Morgan, we met workers whose families have been with Morgan for three generations. Most of these workmen could not imagine working anywhere else.

They value the spirit of the old family business – which is itself in its third generation. They also treasure the freedom they enjoy as they apply their skills at England's last independent, privately-owned automaker. A strict schedule is kept only for the start and end of the workday, as well as the thrice-daily breaks, which are announced in an old-fashioned way by the sound of a handbell. Indeed the individual departments have to meet production schedules, but other than that, each workman proceeds at his own pace.

Seldom have we seen so many satisfied faces in a production facility, to say nothing of the occasional song or whistled tune. Employee turnover is a rare event. New workers usually appear as apprentices, the workforce's youngest members, to learn their craft, often from their own fathers in the same shop.

In this tiny corner of the world, far from the madding crowd, where else would the new blood come from but from the families that have lived there for generations? A factory such as Morgan's is, in many respects, unique in our time and at best comparable to the manufacturers of mechanical watches or musical instruments. But those aren't drivable ...

In this, our third Morgan book, above and beyond all material and technical information, we would like to bring to life the spirit of the company. We are not providing detailed instructions for the assembly of a Morgan – although it might be more cost effective to buy all of the spare parts instead of the completed automobile. However, what individual has a command of all the collected craftsmen's skills and experience embodied within this seasoned team?

Moreover, at Morgan, cars are not simply assembled from individual components. Every part needs to be individually fitted, and, if needed, reworked by skilled hands – mechanical parts as well as the wooden members of the frame and the aluminium skin over that wooden frame.

Thanks to this synthesis of fabrication and manufacturing, customers' individual wishes can be incorporated, if necessary even just before production is initiated. The dilemma of choices in configuring a Morgan begins with the endless variety of customisation possibilities.

These include not only more than 40,000 available paint colours – the limit of the world-wide colour palette – or the roughly 150 different leather colours, but also instrument panels laid out and configured to customer order, the shape of the front end, the bumpers (with or without), the possible deletion of door locks, the type of folding hood ... This list can be extended ad infinitum.

Whether or not the reader possesses a Morgan, or if, after reading this book, you might consider purchasing one, or if you would simply like to be present at the creation of one of these fascinating automobiles – we invite you, via this book, to experience a unique and exceedingly vibrant piece of automotive history.

THE MORGAN STORY
How it all began ... and where it led

Our story begins in 1534. Back then, the framework for the eventual founding of the Morgan marque was established. Isn't this a bit of overreach into the mists of time? Actually, no! In 1534, parliament decided to renounce the authority of the Pope in the Act of Supremacy and appointed the King of England, at the time Henry VIII, as head of the Anglican Church.

This decision led, among other things, to the ability of priests to marry and father legitimate children – and a few centuries later, a certain Henry Frederick Stanley Morgan was born. His father was George Henry Morgan. As prebendary of the congregation of St. Peter and St. Paul in Stoke Lacy (Herefordshire), he had assumed the office in 1887 from his father, Henry Morgan.

Henry Morgan (the second) married Florence, who on August 11, 1881, bore their first child: Henry Frederick Stanley Morgan, known as H.F.S. for short, or Harry. For Morgan fans, the small church has now become a place of pilgrimage. Not only because the Morgan family grave is in the church cemetery, but also because two three-lobed, round, leaded stained glass windows in the porch of the church have a Morgan theme. One shows a portrait of Peter Morgan, Harry's son, a view of the workshops in Pickersleigh Road, Malvern Link, as well as a traditional Morgan (4/4 or Plus 8). The other includes a pair of three-wheelers of differing configurations.

But let's return to H.F.S. He grew up as the third generation of Morgans to live in the sheltered surroundings of a decidedly spacious rectory in the English countryside. This, of course, does not automatically lead to the founding of an automobile factory. But the elder Morgan was not entirely blameless in this development. George Henry Morgan had wide-ranging interests. Among other activities, he was an

George Morgan, c.1890.

Harry's grandparents, Henry and Mary, with George, c.1876.

The Morgan family home in Stoke Lacy, c.1890.

How it all began ... and where it led

George Morgan with relatives, wife Florence, daughter Dorothy, and son Harry, Stoke Lacy, 1892.

Harry's birthplace, the 'New House' – later 'Petersfield' – in Herefordshire.

enthusiastic photographer, with a decidedly gifted, artistic style that garnered multiple awards.

Above all, he had a penchant for the achievements of technology. The subject of automobiles was of very special interest to him. He joined two automobile clubs, advancing to vice-chairman of one club in Hereford.

George Henry Morgan's first car was a Lanchester, acquired in December 1902 – an exclusive vehicle in its day. A respectable percentage of the 3000 miles covered by the Lanchester each year were driven by his son, by that time in his early 20s.

Two other characteristics inherited from his father led to H.F.S.'s development into the founder of an automotive manufactory. First – and in marked contrast to the conventional, strict, patriarchal Victorian upbringing of the time – the elder Morgan allowed his son to choose his own career. Second, he had a knack for handling money. Both traits would have critical importance.

H.F.S. decided to become an engineer. First, he studied at the Crystal Palace School of Engineering in Sydenham, south of London. He completed his studies in the workshops of the Great Western Railway.

Here rests the founder: the Morgan family plot in the Saints Peter and Paul church cemetery.

Today, the church in Stoke Lacy is a pilgrimage destination for Morgan fans.

There, he soon decided that railbound vehicles were not his passion. In 1905, he left GWR and, initially with his partner Leslie Bacon, established an independent garage in Malvern Link servicing Darracq, Wolseley as well as Siddeley automobiles.

So much for the topic of choosing one's own career. Naturally, H.F.S. soon wanted to drive his own automobile. He had a two-cylinder Peugeot engine of 7 horsepower, which he originally intended to put into a motorcycle. In 1908, however, he thought better of it and set to work developing a car.

Of course, it helped tremendously that he was allowed to use the well-equipped workshops of Malvern College, the local establishment of higher education. Morgan owed this fortunate circumstance to his friendship with the brothers Robert Louis and William George Stephenson-Peach, whose father served as chief engineer of the workshops.

A special honour: In Saints Peter and Paul church, colourful stained-glass windows feature a portrait of Peter Morgan, the factory, and a classic Morgan …

… as well as a pair of three-wheelers.

How it all began ... and where it led

Stoke Lacy, 1898: In the background George, to his right Harry, Leslie Bacon in the light-coloured jacket, Florence Morgan at right front, and other relatives.

In 1885, Harry, aged three, posed in a sailor suit.

George in his Lanchester, 1905

Harry sent this Christmas card to his parents in 1903.

Morgan and Co. advertisement in the *Malvern News* of May 27, 1905.

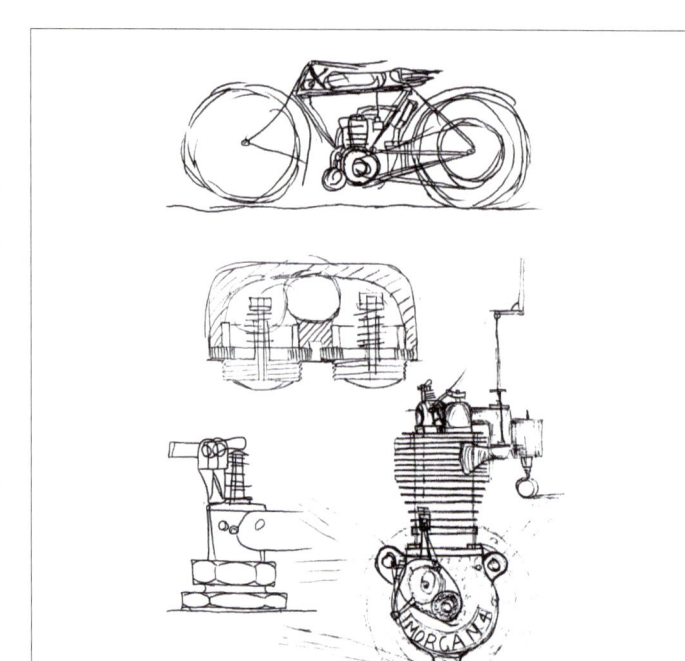

Sketch. How Harry imagined his motorcycle, 1902.

All new. Morgan introduces itself, May 1905.

The showroom, 1905: an eclectic mix of cars under the roof of Morgan & Co.

How it all began ... and where it led

Harry Morgan, c.1907.

Harry (right front) and his 'motor pals' Leslie Bacon (right rear), George (left front) and Robert Stephenson-Peach (left rear), 1905.

The workshops at Malvern College, seen here c.1907, were well-equipped.

THE MORGAN STORY

Morgan three-wheeler

Harry Morgan had to steer the nearly completed prototype of 1910 by means of a tiller arrangement.

Morgan's central idea was to place the engine ahead of the front axle of a three-wheeled vehicle. The result was a cyclecar, or runabout, in the vernacular of the time. H.F.S.'s vehicle had two wheels at the front axle and a single, powered rear wheel. The front wheels were steered by means of a tiller-like arrangement.

The delicate frame, artfully welded, was made of steel tubing, and helped keep the kerb weight below 80 kilograms. This represents specific power (the weight-to-power ratio) of about 11.4kg/hp. Even today, 107 years later, the current Morgan 4/4 has a specific power of around 7kg/hp, despite a kerb weight of 795 kg, an impressive power-to-weight ratio among modern automobiles.

Among the excellent ideas embodied in H.F.S.'s creation was independent front suspension, this at a time when automobile designers were still guided by horse-drawn carriage practice and most cars employed a front beam axle on leaf springs.

Thanks to its independent suspension and outstanding power-to-weight ratio, the car's performance was decidedly sporty. H.F.S., however, cannot claim to have invented independent suspension. That honour goes to Joseph Guédon and Gustave Cornilleau who, in 1897, were the first to mount such to a voiturette, which also happened to be a three-wheeler.

Morgan Three-wheeler

Harry's youngest sister, Dorothy, in the prototype, 1910.

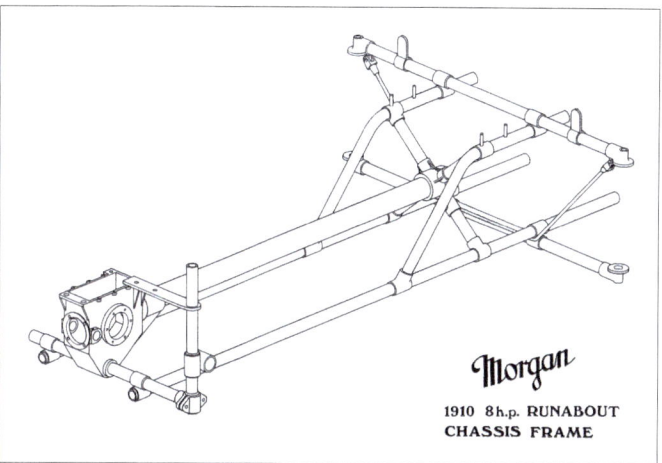

Chassis frame for an 8hp Runabout, 1910.

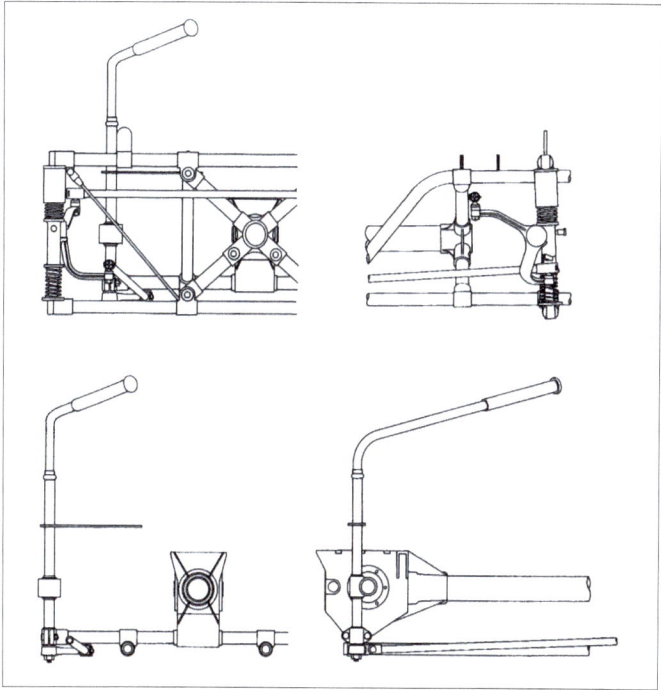

Steering for an 8hp Runabout, 1910.

By 1910, H.F.S. had garnered so much attention with his cyclecar that he let himself be talked into laying on a small limited-production run of similar vehicles. This was the 'starting gun' for the Morgan factory in Malvern Link. Now, his father's second unique character trait came into play – his facility in handling money. Through his extra clerical activities, he had accumulated private assets. As his son had inherited his passion for motorised vehicles, the latter's technical ideas and projects likely gave the father no small measure of pride.

Regardless, the elder Morgan gave H.F.S. the financial means to build up his small original works, new buildings rising alongside the old workshop in Worcester Road, as well as

Pictures from the 1910 Runabout catalogue.

funds for the purchase of the necessary machinery. For several years after its founding, George Henry was responsible for the finances of the Morgan Motor Company, leaving H.F.S. free to dedicate his time to the engineering side.

The fact that the Morgan factory is the sole remaining independent standard-bearer for the once-proud British automobile industry is probably attributable to the fact that the prebendary George Henry Morgan attached a condition to all of his financial dealings: development and future expansion should only be carried out by self-financed means – never on credit.

The resulting financial independence, combined with very conservative investment policies, has certainly been a major influence on the Morgan family's dealings and the image of the factory, all the way to the present day.

One of the greatest investments was made in 1914, when H.F.S. had a new factory erected in Pickersleigh Road, near the old workshop – a factory that remains standing to this day. Even 100 years ago, this facility turned out about 1000 cars per year.

The fundamental features of the first Morgan's front suspension design can still be found today in the marque's traditional models – this in keeping with the principle that 'What has been proven to work at Morgan, won't be changed.' The workshop-based manufacturing process has also remained: in the production process, the car is pushed by hand from one section to the next. There is no assembly line.

Let us return to the beginnings of the Morgan story. After the first vehicles were completed, H.F.S. presented them

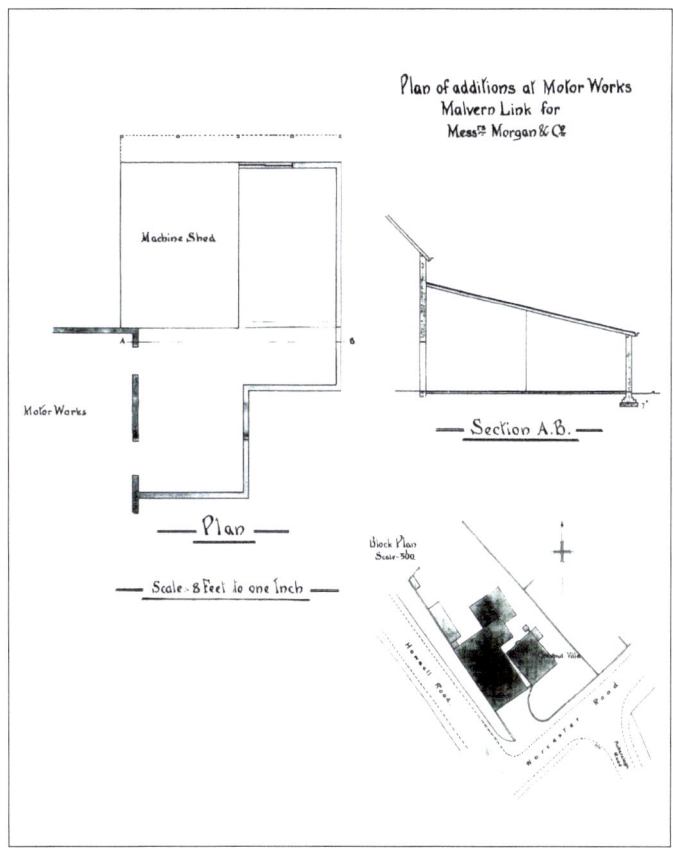

First expansion. Plans for additions to Worcester Road are dated July 20, 1912.

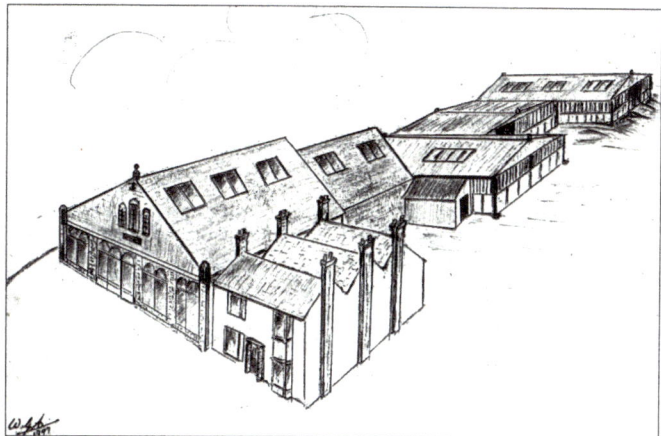

The Worcester Road Factory, 1936.

The Morgan workforce presents the 1913 model line.

The machine shop in Worcester Road, 1910.

to the public at London's Olympia International Motor Cycle Show. Despite good press, the number of orders was less than overwhelming. What was missing was sporting success in so-called 'trials,' in which vehicles could demonstrate their performance and reliability. Harry was able to achieve public attention by participating in the wintertime London-Exeter Trial, among other successes.

A factor causing some sales resistance was the fact that the first model produced was a single-seater, limiting its utility. To remedy this, in 1911 H.F.S. developed a two-seater and replaced the tiller with a steering wheel. He also secured a prominent retail partner in the famous and innovative London department store Harrods, in London's Knightsbridge district. Harrods presented the Runabout in its showrooms as a specialty item, which resulted in an appreciable boost in sales.

Morgan grows. The new paint shop and assembly hall in Pickersleigh Road, 1914.

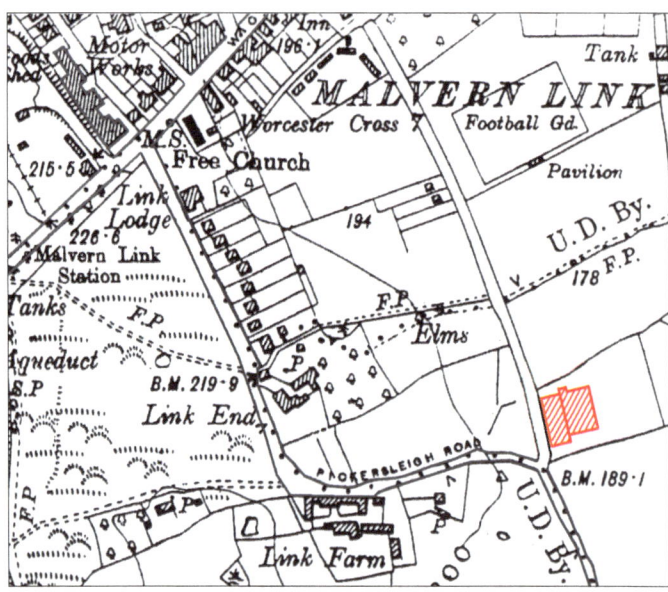

This plan, dated 1914, shows the arrangement of the new structures along Pickersleigh Road.

Morgan began adding sporting successes in 1912. At its premiere in the Brooklands cyclecar race of that year, a Morgan scored a victory for the marque in its very first attempt.

Even more important than such victories was the average speed record, at the time regarded as the sole yardstick for performance. As nearly every maker of cyclecars strove to break the record set by H.F.S. in 1912 at 57 miles per hour, Morgan's record did not stand for long.

With dogged determination, H.F.S. worked on improvements to crack the 60mph mark. In November 1912, although he fell short of the magic distance by only a few yards, he did recapture the record for distance covered in an hour; the Morgan was once again the fastest cyclecar in existence. The year 1912 also saw H.F.S. marry Ruth Day, the daughter of a vicar in Malvern Link.

The success of three-wheeled cyclecars in England was not only based on their sporting successes, or the fact that such vehicles could run rings around many four-wheeled automobiles. The primary consideration was economy. In their most affordable version, the first Morgans were priced at just £65. Even at the time, that was a very moderate price.

'Why not a Trio-Car?' Concept sketch by John Bryan, 1910.

Harry Morgan at the tiller, with brother-in-law William Cowpland in the two-seater prototype, 1911.

The Harrods Morgan Runabout was presented in December 1911.

Added to this, motor vehicles with fewer than four wheels – regardless of engine displacement or power – were taxed at a rate of just £1 per year. This exceedingly public-spirited tax rate remained in effect until 1920. Of course, taxes were raised in later years, but until Morgan ceased production of the models in 1952, a three-wheeler weighing less than 500 kilograms cost a flat rate of £10 per year in taxes, whilst drivers of four-wheeled cars were taxed on the basis of horsepower.

Morgan achieved its first international racing successes in 1913. H.F.S. prepared a total of four vehicles, fitted with various water-cooled engines, for the *Cyclecar* Grand Prix in Amiens, France. In an exciting race, punctuated by mechanical breakdowns and retirements, W.G. McMinnies and his riding mechanic Frank Thomas were first across the finish line in their J.A.P.-powered Morgan. The year 1913 would see further Morgan victories.

Those days also saw the first exports of Morgans to France, establishing a beachhead on the Continent. The reputation of the marque, tied as it was to racing successes, led to such brisk sales that by the end of the year, the entire production run for 1914 had been pre-sold!

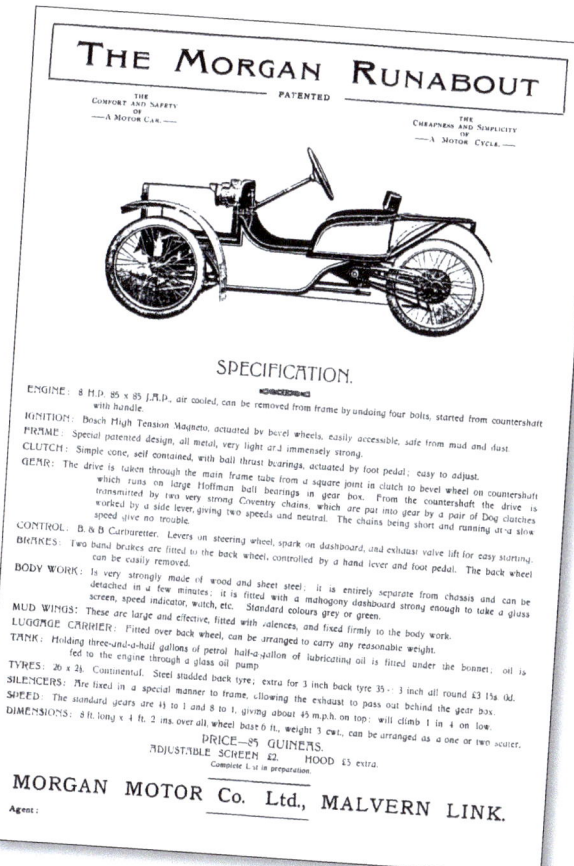
Now with steering wheel!
The Runabout for the 1910 Olympia show was priced at 85 guineas. An adjustable screen was a £2 option, the hood another £5.

THE MORGAN STORY

June 17, 1907 was the very first race day at Brooklands.

Harry Martin at the start of a race, Brooklands, 1912.

All in white: Ruth and Harry on their wedding day, 1912. All matching the magnificent wedding carriage.

On the cover of *The Cyclecar*: Harry in the record car, George (with top hat), an official timer (with bowler hat) and, behind them, Ruth.

Win on Sunday, sell on Monday – Harry's Brooklands win and one-hour record on March 27, 1912 were immediately featured in advertisements.

Despite the outbreak of World War I in 1914, H.F.S. continued to develop new models and set new speed records. Although the Morgan factory produced mostly munitions during the war, cyclecars continued to be made in limited numbers. This activity was approved solely for export purposes. The cars were shipped to Canada, Bolivia, India and Russia.

As soon as the war ended, production was once again cranked up, as economical vehicles were in especially high demand due to fuel shortages.

The fact that Morgans could be fitted with any of a variety of very different engines also proved fortuitous as, after

THE MORGAN STORY

The start of an exciting race, punctuated by many breakdowns.

The leap to the Continent: a win at Amiens for W.G. McMinnies and his mechanic Frank Thomas, aboard a J.A.P.-powered Morgan.

the war, no single engine supplier could be relied upon to deliver in the desired quantity.

Just months after the end of the war, fuel supplies returned to normal, and the wartime ban on trials was lifted. Naturally, Morgan drivers once again took their places on the victory podium. The new workshops in Pickersleigh Road, built in 1914, were soon turning out up to 2500 vehicles per year.

There were also changes in the Morgan family. On November 3, 1919, Peter, the first of five children born to Harry and Ruth, first saw the light of day – born in the Morgan home directly next to the factory.

The 1920s saw a broad model range, with two- and four-seaters, available with or without folding hoods, and

The Morgan family with their cars in Stoke Lacy, spring, 1914.

even a delivery van model. Still, slowly but inevitably, the cyclecar era was nearing its end.

Even racing successes and speed records – in 1930, Gwenda Stewart, the fastest woman in a Morgan, drove an 1100 cc model to 117mph (186kph) – could not avert the demise of the three-wheelers. The automotive future rolled on four wheels. By 1935, sales of cyclecars had dropped to 286 examples per year.

Despite the principle that nothing proven to work would be changed, H.F.S. developed the Morgan 4/4 – a vehicle with four cylinders as well as a four-wheeled chassis. In doing so, he carried over, almost without change, the front suspension of the first three-wheeler of 1910. In principle, this suspension system exists to this day, in all of Morgan's traditional models.

In the Autumn of 1918, Harry and Ruth showed just how much could be fitted into the pre-production family model.

The Morgan family, with baby Peter.

The Morgan factory in Pickersleigh Road, northwest view: 1920.

The brand-new Grand Prix Morgan has just been completed, 1922.

Morgan Plus 4 in front of the factory gate, 2015.

Morgan 4/4 and Plus 4

A classic: Morgan 4/4, 1936.

The 4/4 was officially announced on December 27, 1935 and, in typical H.F.S. style, immediately entered in a trial, and driven to victory by H.F.S himself. Still, the three-wheelers remained in production. Their era did not end until July 29, 1952, when the last three-wheeler left the Pickersleigh Road works. Until then, three-wheelers continued to compete in trials and rallies. At the 1938 Land's End Trial, for example, seventeen Morgans were entered, seven of these three-wheelers.

Prebendary George Henry Morgan passed away in 1937. H.F.S. succeeded his late father as company chairman and governing director. H.F.S.'s wife, Ruth, also served on the company board. To head day-to-day operations, H.F.S. promoted long-serving works director George Goodall to managing director. At this point H.F.S. not only changed residences, but also withdrew from daily operations, instead visiting the works several times per week.

The following year saw a major motorsport success. In 1938, a remarkable woman took part in the 24 Hours of Le Mans. Miss Prudence Fawcett entered and drove a Morgan 4/4. Accompanied by her co-driver Geoff White, she finished 13th in a field of 42 cars – even though many of these had much larger engines, displacing up to four litres. In her class (below 1100 cc), Miss Fawcett took fourth place and, for many years, this remained Morgan's best Le Mans finish which was not topped until 1962. But more about that later.

Until 1939, the Morgan 4/4s were fitted with Coventry Climax engines, as Morgan remained true to his principle of not developing his own powerplants. The body consisted of steel sheetmetal, mounted on an ash frame. This wooden frame survives to this day on the two- and four-seat models. The doors were all hinged at the front – a practice that continues on Morgans to this day. The Coventry Climax engine displaced 1122cc and proudly produced 34 horsepower.

The company did not weather the second World War as easily as it had the first. The Morgan factory was occupied by another firm that built fuselages for Avro Lancaster bombers. After the war, there were numerous problems. Production could not be resumed as quickly as in 1918,

George Morgan in the early 1930s.

Morgan Plus 4 Drophead Coupé, 1959.

TOK 258, the famed Le Mans class winner of 1962, returned to the Sarthe for the Le Mans Classics of 2012, with Keith Ahlers at the wheel.

after World War I. The production facilities were long occupied by other firms, and workers only gradually returned from their military service.

Added to these problems were difficulties with suppliers. Raw materials, above all steel, were in short supply. Only the firm's strong emphasis on exports shielded Morgan from even greater damage. Export was the only means of accessing Britain's post-war steel rations in order to slowly rebuild production.

It was not until early 1949 that a measure of normality became apparent, as fuel rationing was relaxed and the first post-war motorsports events took place. New models, such as the elegant Drophead Coupé, were created.

In 1951, the successor to the 4/4 – the Morgan Plus 4 –was introduced. Visually, the Plus 4 differed little from the 4/4; the major change was the engine. It was already known that intended supplier, the Standard Motor Co., would cease production of its smaller 1.2 litre engines, Morgan developed the Plus 4 to take the 2.1 litre powerplant, which remained in the model line from 1951 until 1958.

The Plus 4 is a veritable survivor, built with few interruptions from 1951 to the present day. Only engine suppliers changed.

This Morgan Plus 4 Super Sport of 1954 took part in the 2014 Le Mans Classics.

OAK Morgan LMP2 racer at the 2013 Bahrain 6 Hours.

Some engine variants were installed successively, while some left the workshops in parallel with other motors.

Power units included the Standard Vanguard (1951-1958), Triumph TR2 (1954-1956), TR3 (1956-1962), TR4/4A (1962-1969), Fiat (1985-1987), Rover M16 (1988-1992), Rover T16 (1992-2000). Currently, newly built Morgans are fitted with Ford engines.

Now and again, Morgan would produce limited numbers of so-called 'Super Sports' models, which caught attention on the world's race tracks. For example, in 1962, Chris Lawrence and Richard Shepherd-Barron took second place at the Spa

Morgan Plus 4 of 2011.

Grand Prix. Whereupon they drove to Le Mans and took part in the 24 Hours race. The duo not only improved on Prudence Fawcett's drive, 24 years earlier, but also won the under-2-litre class and finished 13th overall.

This historic Morgan victory has not been repeated to this day – apart from the win by the OAK Morgan Nissan that competed in the LMP2 class in 2013. That car was built in its entirety by OAK Racing in Le Mans.

Apart from engines, the technology of the traditional Morgan models hardly changed up to the time of writing. It could be said that what Morgan called a "major change" at the end of 1953 was, in retrospect, a facelift. Changes at that time mainly affected the front aspect: the free-standing headlamps were integrated into the bodywork and the radiator grille was curved. These changes gave the car a somewhat more modern look.

If one compares a 1953 Morgan with its sports car contemporaries, one might be less inclined to apply the term 'modern.' Compared to the English classic, the early Corvette of that year and the Mercedes 300 SL, introduced a year later, were more like props from a science fiction movie.

For his part, Peter Morgan, who joined the firm in 1947 and officially assumed its leadership in 1959, was less than enthusiastic about the new design. He felt it would not be especially long-lived. In that prediction, he completely missed the mark; the shape continues in production to this day.

Over time, a few minor touches were applied to the bodywork. Above all, the wing widths were altered to accommodate ever-wider tyres. Doors and interior space were enlarged ever so slightly. Externally, the car's overall length grew from 3.66 metres to 4.01 metres (12ft to 13.45ft). The two spare wheels at the back, attributable

The 2015 4/4.

to the frequent flat tyres of the time, were at some point eliminated entirely, and the interior details underwent occasional, minimal modification.

In 1955, the smaller 4/4 experienced a renaissance. It was now equipped with various Ford powerplants ranging from the Ford 100E engine, displacing 1172cc and making 36 horsepower, to today's Ford Sigma engine, with 1600cc and 112bhp. For a brief period, between 1981 and 1985, Fiat engines were installed. In 2015, the 4/4 remains in the model line up.

Except for engines, the 4/4's construction is identical to that of the Plus 4. In 1998, the previously typical steel body, draped over a wooden frame of ash, was replaced by an aluminium skin. Initially, the car had front and rear drum brakes; disc brakes (at least, at the front) were available as an option from 1959, and became standard in 1961. Rear drums have remained to this day.

Original cars had four-speed transmissions, finally replaced by five-speed gearboxes in 1985. In the current incarnation, these are supplied by Mazda. Also persisting to the present are the characteristic grease nipples on the front suspension.

Since the vehicles have remained basically unchanged since the first four-wheeled Morgan of 1936, one could describe these Morgans as, essentially, pre-war cars. In a Morgan, one is actually driving a vintage car whilst meeting modern exhaust emissions standards.

We are talking about a veritable veteran motorcar. Beyond the aforementioned changes to engine, transmission, and interior appointments, neither the 4/4 nor the Plus 4 are available with ABS, ESP, or airbags – in other words, nothing that isn't absolutely required for the act of driving.

Morgan Plus 4 Plus

The Plus 4 Plus was a beautiful failure.

In 1963, Morgan underwent something akin to a revolution. For the first time since the Morgan 4/4 of 1936, a completely new vehicle concept appeared wearing the Morgan badge. The Plus 4 Plus was presented, with great hopes, at the London Motor Show. It was intended to be the Morgan of the future: a modern, lightweight automobile, powered by a Triumph TR4 engine.

It was the first Morgan in history to have a proper luggage compartment in addition to the storage area behind the front seats. It was a vehicle suitable for long journeys – a Grand Tourer – elegantly dressed in modern fibreglass bodywork. In principle, the decision to build such a car was perfectly understandable. However, compared to the competition – in particular, MG and Lotus – the Morgan appeared dated.

What Peter Morgan had not anticipated was the traditional mindset of the Morgan fan community. While Morgan once again enjoyed a successful racing season, and three Morgans swept the podium at Spa in their displacement class (2.0 to 2.6 litres), and the 4/4 and Plus 4 showed halfway decent sales, the Plus 4 Plus was an utter marketplace flop. Until production ceased in 1967, only 26 examples were made, of which ten were shipped to the United States.

Morgan Plus 8

Peter and Charles Morgan entered the Plus 8 prototype in races.

Fast on the hills. The owner of this 1975 Plus 8 took part in the 2015 Prescott Speed Hill Climb.

In 1968, the failure that was the Plus 4 Plus was followed by the factory's all-time highlight. On the night of 16/17 February of that year, a deep-throated rumble emanated from the Pickersleigh Road workshops – a sound that could never come from a four-cylinder engine. Morgan's legendary Plus 8 was born. Over time, this developed into the synonym for sporting performance and power.

The light-alloy eight-cylinder engine – originally designed by Buick, and modified by Rover – displaced 3.5 litres and, at its debut, put out 160 horsepower. More important, though, was its torque of 285 Newton-metres (210lb/ft). Combined with the Plus 8's kerb weight of 899 kilograms, the engine had an easy job of it.

Even though top speed just barely scratched at the 125mph barrier, the Plus 8's acceleration from a standstill to 62mph in 6.7 seconds was sensational. As such, it could even run away from the concurrent Jaguar E, Series 1, with its 3.8 litre engine. In its most powerful version, a 4.8 litre V8, available as an option from 1997 to the end of production in 2004, the Plus 8 had an engine developing 220bhp and 260lb/ft of torque.

With 6222 examples built, the Plus 8 was by far the marque's most successful model. The reason for terminating the Plus 8 was, once again, the end of engine production by the supplier, this time Rover.

THE MORGAN STORY

Morgan Roadster

The Roadster, seen here with the contemporary 3 Wheeler, is fast and powerful.

The Morgan Roadster has a pure and classical presence.

In 2004, as a successor to the Plus 8, Morgan presented the Roadster. Although fitted with a more powerful engine, blessed with 'green' certification, exhibiting better roadholding thanks to lower, more rearward mounting of the somewhat compact three-litre Ford V6, and able to reach a higher top speed with quicker acceleration, in fact able to do everything better than the Plus 8, emotionally the Roadster was never able to take the Plus 8's place.

Roadster factoids: 3.0 litre displacement, 24 valves, 231bhp, kerb weight 938 kg. This resulted in acceleration from zero to 62mph in just 4.9 seconds, and a top speed of 143mph.

In 2012, Morgan started installing the 3.7 litre engine of the Ford Mustang. According to factory figures, this developed a respectable 284bhp, but it was not at all unusual for a particularly healthy specimen to pull more than 300bhp on the dynamometer. Currently, the car uses a six-speed manual transmission, and kerb weight has risen ever so slightly from 938 to 950 kilograms. Other than that, everything remains as it was.

Well, almost everything. Finally, the driver does get a smidgen of outside help. Because the heavy engine results in heavier steering, Morgan decided to allow the Roadster an electric power steering system. Naturally, this remained a Roadster exclusive. Drivers of the 4/4 and Plus 4 continue to steer with human muscles alone.

Morgan Aero 8

Let's go back a few years. In 1999, the third generation of Morgans, Charles, son of Peter Morgan, took on the mantle of the family firm's fortunes. Since the mid-1990s, Charles Morgan had been thinking about the Morgan of the future – a completely different Morgan for the next millennium. He was convinced that if Morgan intended to remain in the major leagues of sports cars, one would have to think about matters beyond mere engine choices.

Above all, his plans involved the chassis and suspension. It was no secret that the classic Morgan chassis dated back to 1936, and was in principle a further development of the very first three-wheeler into a four-wheeled vehicle. At least for motorsports purposes, this solution could not hold sway forever.

Peter and Charles Morgan with a Plus 8.

As was Morgan practice from the beginning, the new car was unveiled in Geneva. Thirty-seven years after the Plus 4, Morgan presented the Aero 8 at the 2000 Geneva Auto Show – another super-modern car (as the house of Morgan defines super-modern).

The Aero 8 remains a polarising vehicle, thoroughly contemporary, yet in its construction a genuine Morgan: aluminium chassis, an exterior skin of the same material, and – how could it be any different – a frame of English ash.

Morgan Aero 8.

And now for something completely different: the new boot of the later Aero 8.

For propulsion, this time Morgan helped itself to the BMW parts bin: a V8 displacing 4.4 litres, and good for 286bhp, provided the traditionally lightweight car with 0-62 mph acceleration in less than five seconds. Despite an air conditioning system and power windows, the Aero 8 was nevertheless a purist's car: ABS, ESP, and airbags? Forget it – at least in Series 1 examples ...

Furthermore, the headlamps were placed in a way that gave the model a cross-eyed gaze.

The Aero 8 was nevertheless a good platform for motorsports. A special series, the GTN, was conceived for the British GT Cup and for appearances in the Le Mans 24 Hours. Right off the bat, it took second and third places in the 2002 GT Cup. For the roadgoing Aero 8, output of the bored-out 4.6 litre V8 was pushed to 330bhp. Eleven cars with rigidly mounted hard tops, contrasting silver and blue paintwork, numerous carbon fibre components, semi-slick tyres and racing suspension were licensed for the public road.

At the 2004 Los Angeles Auto Show, Morgan presented the Series 2, exhibiting considerable engineering improvements. The 4.4 litre V8 now developed 333bhp, a new transmission by ZF provided smoother gear changes. Morgan's new meteor was given larger six-piston caliper disc brakes and,

The Aero 8 formed a good foundation for motorsports, for example in its GTN incarnation.

amazingly, electronic braking and traction assist – both firsts for Morgan.

This, the safest Morgan to date, could even be registered in the USA. There was one major exterior change for the Series 2: The smooth, flowing rear end was restyled. The Aero's resulting prominent bustle back remains a topic of discussion and dissent. Beginning with the new model, customers could choose between the classic rear exhaust or side pipes, which became even more popular in the later Series 3 and 4 which had room at the back for an optional underbody diffuser.

In 2006, after only 80 examples of the Series 2 had been sold, Charles Morgan presented its successor. A successful 'eye operation' cured the patient of the cross-eyed look,

but the rear bustle remained part of the Aero styling until production ended in 2009.

The final model was the Series 4, unveiled in late 2007 with an even more powerful engine. This produced 367bhp and 369lb/ft of torque. For the first time, a Morgan could be ordered with an automatic transmission. The ZF six-speed automatic was an outstanding match to the powerful engine, and gave even quicker acceleration than the manual-shift version. Top speed was now an astonishing 170mph.

There was a technological revolution in the offing. To supply the now not insignificant number of electrical systems in the car, the Series 4 was given a CAN bus, a sort of data superhighway for data information exchange between

The Aero 8's characteristic 'ducktail' remains controversial to this day.

the car's systems. After ten years and about 700 examples, production of the Aero 8 ended in late 2009.

Again in keeping with tradition, a successor bearing the same name was presented at Geneva in 2015. With the new Aero 8, Morgan sought to cultivate its tradition. The fifth iteration is a further development of the familiar Aero models. The chassis is reworked, and there is a new, more rigid front suspension design with an anti-roll bar. The rear suspension is equipped with a modern BMW limited-slip differential.

Oh, and one more thing: the tail had been completely reshaped and no longer carried the bustle. For the first time, as an option, buyers could order an 'infotainment system' with integral navigation system meaning that, for the first time in this vehicle class, owners could do away with those suction cup gadgets on their windscreens.

The Aeromax is a modern interpretation of the Morgan theme.

In February 2005, a young design student named Matthew Humphries, who had completed an apprenticeship in the Morgan works, was responsible for that year's biggest head-turner at the Geneva show. In its technology, the Aeromax was based on the Aero 8. Intended as a unique one-off, built for a friend of the Morgan family, it turned into a limited series of 100 examples. Intentional or not, the car was a homage to the legendary Bugatti Type 57 SC Atlantic, and a true feast for the eyes. Matthew was quickly hired as Morgan's chief designer, and stayed with the firm until 2013.

Aero Supersports/Aero Coupé

The Aero Supersports, seen here in the surroundings of its historical factory, might be likened to an art deco spaceship.

Production of the old Aero 8 was terminated in 2009, only to be immediately replaced by a new model, available as a Targa and as a Coupé. Right on time for Morgan's 100th anniversary, and in the presence of Princess Anne, the anniversary model – the Morgan Aero Supersports – celebrated its debut at the Concorso d'Eleganza Villa d'Este on the shores of Lake Como. With a powerplant befitting its status – a BMW 4.8 litre V8 of 367bhp – the Aero Supersports could accelerate from a standstill to 62mph in 4.5 seconds.

The Aero Supersports and Coupé appear simultaneously classic and fresh.

Morgan Plus 8/Speedster

Wearing a classic track suit and trendy running shoes: Morgan Plus 8.

No, that's not a typo. This is a new chapter. In 2012, eight years after production ended, a Plus 8 was once again presented. Like the other eight-cylinder models whose drivetrain it shared, this re-interpretation of the successful model was based on the Aero platform.

The shape, however, was strongly reminiscent of the traditional model. Its overall length was identical to that of the 4/4 and Plus 4, to the millimetre; only its width had increased, a fact attributable to the choice of platform. With a kerb weight of only 1100 kilograms, it was the world's lightest V8 – until the Speedster was introduced in 2014.

There's always something to celebrate in the lively Morgan workshops. Most recently, in 2014, it was the centennial of Morgan's manufacturing activities in Malvern Link's Pickersleigh Road. Because there were still a few hundred BMW engines in the warehouse, and these needed to be moved on to customers, it made sense to create another special edition – in this case, the Speedster.

The intent was to build just sixty examples, with sequentially numbered plaques mounted in the interior. Equipment was reduced to the basic driving necessities. A windscreen? What on Earth for? A diminutive plastic windscreen – a sort of Brooklands screen – would suffice. Windscreen wipers? For what windscreen? Sidescreens? Bah! Useless ballast! A hood? To which windscreen frame would that attach? Instead, a tonneau cover was included as standard equipment.

The quantifiable benefit of this orgy of deletion: a few kilos less kerb weight and a price of around £65,000 – for an

Geneva, 2015: Morgan's then latest Plus 8 on display at the annual International Geneva Motor Show.

eight-cylinder Morgan, a real bargain. And for all those who would rather not do without some weather protection, there was a retrofit package consisting of a windscreen, sidescreens, windscreen wipers ... and a hood.

The Speedster sold well. This gave the men of Malvern Link the idea of turning the special edition into a regular model. In 2015, at the 85th Geneva Auto Show, Morgan presented the new Plus 8 generation: no windscreen, no wipers, no sidescreens, no hood ... wait a minute, didn't there used to be a special edition like that, called the Speedster?

3 Wheeler

Race ready: a new 3 Wheeler making rapid progress at the 2015 Prescott Speed Hill Climb.

When production of the three-wheeler ended in 1952, it was already regarded as a rolling anachronism from the Edwardian Age, prior to World War I. Yet not quite sixty years later, in March 2011, a brand-new three-wheeler stood inside the Malvern Link workshops.

The honour for this rebirth actually goes to an American. Pete Larson of Seattle had bought a license from Morgan to produce cyclecars, and powered by Harley-Davidson engines, sold a small number of these to the public. The winged ACE logo was strongly reminiscent of the Morgan badge.

Morgan managing director Steve Morris heard of the retro 3 Wheeler and in 2010 flew to the States to have a firsthand look. Morris bought an example and brought it back to England. At Malvern Link, the engineers disassembled the cyclecar into its component parts and considered whether or not a revival of the 3 Wheeler line might make sense for Morgan.

Fortunately, their decision was in the affirmative. The license was bought back, and in short order Pete Larson was contracted to be the US agent for the Morgan three-wheeler.

The call of the open road: prototype of the new 3 Wheeler, as seen in the factory in 2011.

In addition, Pete Larson also builds motorcycle sidecars, under the 'Liberty' label.

In 2011, the new three-wheeler was displayed at the Geneva salon – not yet roadworthy, but visitors had no way of knowing that. To everyone's surprise, at the end of the auto show, Morgan had 300 orders in hand. A short time later, as we researched factory records for our first Morgan book, we saw two lonely 3 Wheelers in an otherwise nearly empty factory hall. Production had not yet started, but a few wooden frames of English ash could already be seen on the warehouse shelves. And there was enormous pressure from dealers who wanted to finally deliver the ordered cars to their customers.

Three years later – 2014 – 3 Wheelers accounted for just over 25 percent of Morgan production. At first, their share of production was even higher, but once the backlog of orders had been cleared, the initial output of twelve cars per week was dialled back to three to five per week.

This modern cyclecar is powered by a two-litre S&S V-twin, originally developing 115 horsepower, but reduced to 82bhp for production. This is enough to accelerate these cars from a standstill to 62mph in six seconds.

3 Wheeler test drive

For easier entry to the vehicle, the steering wheel can be removed, just as on a Formula 1 race car. In the snug cockpit, the driver feels like the pilot of a World War I fighter plane. The view of the road is over tiny windscreens, vaguely reminiscent of the old Brooklands racing windscreens.

The starter button is secured under a protective cap. So … flip up the cap, depress the clutch pedal, and press the starter button. The mighty V-twin springs to life and shakes the sleep from its bones. It quickly settles down to a steady idle at just under 1000rpm.

Operating the clutch requires well-developed calf muscles, but the gearbox, from the Mazda parts bins, rewards with light, precise, short throws. Road manners are impeccable, even as the limit of adhesion of the tyres tests the driver's bravery. It is easy to see how the three-wheelers of old behaved on the race track and in trials.

And so the circle closes. Then as now, the 3 Wheeler is a successful model, albeit for completely different reasons in today's market. It is a fun car par excellence – but unfortunately no longer subsidised by the tax man. Not even in England.

What's in a word –
James Gilbert on the topic of sustainability

"Even though it may sound a bit trite, environmental protection and sustainability are topics to which Morgan has been dedicated for years, even at a time before those words were seldom or never used – indeed, long before there was ever such a concept as 'sustainability'. Because these buzzwords are now in common use, and purchasing decisions are based on them, we naturally don't want to miss the chance to reflect on them here.

It has long been known that the frame of a Morgan is made of English ash – that is, of a proven renewable resource. It is also known that the raw material suppliers must be obligated to plant more new trees than are needed for production. This has the noticeable effect: England, and specifically the Midlands, would probably have noticeably more forests if Morgan were to significantly expand its production.

The important point, however, is that a Morgan does not end up in a scrapyard at the end of a defined life cycle, but rather is built for eternity. A Morgan is eventually sold or inherited – at least, this is true of the four-wheeled vehicles delivered since 1936, which have long enjoyed cult status among enthusiasts.

The first three-wheelers fulfilled their mission as affordable everyday automobiles, and were produced in large numbers for their time. Of those, few examples have survived, but those few are well cared for and meticulously restored – also for the benefit of generations to come.

Of all the four-wheeled Morgans ever built, it is estimated that more than 90 percent are still operating on this planet's roads. And that percentage is rising, since anti-corrosion measures as well as manufacturing quality have been greatly improved.

Thanks to the large proportion of hand craftsmanship involved, energy consumption for the production process is rather modest. All of the new Morgans meet Euro 6 emissions standards and, thanks to their light weight, consume comparatively little fossil fuel. At least, this is the case for the 3 wheeler, the 4/4 and the Plus 4. According to a study at the University of Cardiff on the subject of sustainability in the automobile industry, the Morgan 4/4 achieved a veritable first place ranking.

Incidentally, even the Morgan factory is affected by the topics of fleet fuel economy and CO^2 taxes. At least these questions have, for years, prompted Morgan to experiment with alternative power sources based on hydrogen or electrical energy – even if those concepts are not yet available in the marketplace."

Day 1

MAKING A MORGAN

Pickersleigh Road, 2015.

The calendar reads Monday, May 4, 2015. A beautiful spring morning. The English breakfast at our temporary home, the Foley Arms in Great Malvern, is absolutely delicious, and the writing team is eagerly looking forward to an eventful day. The sun climbs into an unusually clear blue sky. Naturally, we put down the hood of our own Morgan – even if it's less than two miles from the hotel parking lot to Pickersleigh Road, home of the Morgan works for more than a century.

We still don't know what awaits us. So far, we don't even know the specifications of 'our' Morgan, whose genesis we will be allowed to witness. We'll just let it come as a surprise. We drive through the historic factory gates, past the legendary factory halls, to the visitors' parking lot. It's not long before we're greeted by our first contact. James Gilbert ushers us into the cafeteria and quickly briefs us on company customs. James is responsible for Press and Social Media at Morgan. All has been prepared for our work, but he asks us to be patient as he has to sit in on the daily production conference.

Where it all begins: Chassis Shop

A short time later, he rejoins us and guides us to the chassis shop, the starting point for all four-wheeled Morgans. It is here that the foundation of the car is built – the so-called rolling chassis. James introduces us to John, foreman of the shop, and shows us the first fragments of 'our' Morgan, order number 92183. From this moment forward, we will witness its production, from the very first to the very last operation.

And so the tour begins. James makes our introductions in the Chassis Shop.

Order number 92183: James goes through the list of details for our customer's car.

Specifications

We will accompany a very traditional model: a Morgan Plus 4 with a 'narrow frame front.' The unique thing about this car is that the front end properly belongs to a 4/4, that is, a 1.6 litre standard model, but the car will have the more powerful 2-litre engine.

Externally, the difference is invisible except to Morgan cognoscenti: On the narrow frame model, the distance between the headlamps and the radiator grille is somewhat less than on the Plus 4 models. Most likely, the new owner prefers understatement.

On the other hand, he or she didn't hold back when specifying the car's equipment: sports model without spare wheel, polished wire-spoke wheels, door pockets, full leather interior in MH Red, map light, hood, sidescreens, tonneau cover in black mohair, so-called front and rear overriders on the bumpers, sun visors, and a radio with CD player.

Oh, yes, before we forget, this car will be painted in the very traditional, dark, Irish Green paint, code L 4069. Why there should be so many green Morgans is a mystery to us; after all, customers can choose from the entire worldwide palette of colours, and this numbers about 40,000.

Although the car will be delivered to a London Morgan dealership, what we have here is a left hand drive model. Most likely, the customer lives on the Continent. He or she doesn't know that a book is being written about their future Morgan. Of course, we will see to it that he will get an autographed copy of this book.

MAKING A MORGAN · DAY 1

Our story so far ... Conversation with Nick Baker on what happens after an order is received

Nick is Morgan's UK Sales Manager, and has been with the company for five years. He explains what happens when a new order is received from one of the 50 worldwide Morgan dealerships.

Every dealer establishes his own annual model-specific sales plan, and reserves so-called production slots for the following calendar year, along with estimated delivery dates. So for example a Plus 4 may be scheduled to begin production in the second week of May.

An order number is sufficient to start the process (in our case, it's 92183). No other specifications are needed at this point. A down payment, however, is due immediately. If a customer has a special wish regarding production or completion deadlines – for example, on a milestone birthday – dealers may trade production slots among themselves.

Where all the parts kits come from: the 'goods in' warehouse.

In the "goods in" warehouse, left- and right-side windows await assembly into sidescreens.

As soon as a customer decides to order a car from his franchised Morgan dealer – after thorough consultation and a test drive – the individual specifications and desired start of production are determined as precisely as possible. The dealer transmits this information through the Morgan intranet and orders the car. This order results in a build ticket, and simultaneously a printout of the bill of material which includes everything the customer would like to see in his Morgan.

At the beginning of production, this list first goes to the ‚goods in – goods out' warehouse, where eager hands under

Conversation with Nick Baker on what happens after an order is received

The green trolley carries the parts intended for the trim shop, including the steering wheel, headlamp trim rings, and radiator.

Steering columns, shock absorbers, and other suspension components are assembled into chassis kits for the chassis shop.

foreman Stuart Webb pack the required materials for each workplace in systematised boxes, which are then distributed accordingly – of course, these are hand-carried.

There is no set time for this task, because the search for specific special-order parts is often very time-intensive. One copy of the bill of material goes to each of the involved workshops. One copy goes to Aero Racing, a legally separate company operating under the roof of the Morgan factory and responsible for special components that the Morgan plant, as manufacturer, does not have – or is not permitted to have – in its inventory, such as wooden steering wheels with non-impact-absorbing hubs.

If a customer decides on an equipment change just before the start of production, this is still possible, for an added fee, until just one day before production gets underway. Of course, it is possible that special wishes will result in longer delivery times. This may be the case for rare leather types, where the factory may have to take delivery of a certain minimum quantity. For paint colours, though, this is not a problem; these are mixed in the paint shop immediately prior to application.

Steve Morris – the chieftain

Steve Morris has served as managing director of Morgan since early 2013. He assumed these duties from Charles Morgan at the behest of the stockholders. He is, however, no newcomer to the company, having begun there as an apprentice in the sheet metal shop.

The fact that he landed at Morgan at all is due to his father, who on several occasions delivered him to the factory gates and told him to find himself a job. The quest was third time lucky. Steve hails from Welland, near Malvern. He is married and father of four, ranging in age from 15 to 23 – and his oldest son works as a CAD specialist in the Morgan Design Center.

Steve sees his mission as largely one of preparing Morgan for the future. He has seen some of his visions become reality: the machinery pool has been carefully brought up to modern standards, there is a continuous effort to improve quality, dealer communications have been greatly improved, and a training academy for the next generation of craftsmen has been called into life.

That last item was urgently needed, as a number of employees will leave the company in the coming years, and young workers with the necessary qualifications are almost impossible to find. For that reason, in the short term, the number of apprentices will be increased from ten to twenty, and a special trainee program implemented.

Even if it really doesn't fit with Morgan's image or history, changing laws, such as introduction of CO_2 taxes in some countries, or establishment of fleet fuel economy standards, force even manufacturers such as Morgan to come to grips with alternative propulsion systems. First, the engineers experimented with an 'e-three-wheeler,' after an e-Morgan based on the Plus 8 had been presented at the 2012 Geneva Show. Modern times indeed at Morgan!

Brett Burbeck – the perfectionist

Brett is 45 years old and hails from Malvern. He has been working at Morgan for 15 years, and always at the same station. Over those years he has developed a remarkable routine. And he is astoundingly well organised. Every hand movement is precise, calculated, effective. All the while he radiates calm and supreme competence. In the company-internal evaluation on his department's 'white board,' Brett outdistances all of his colleagues, with 99 percent efficiency. A glance at his tool cart shows how orderly, indeed almost pedantically he works. Everything is not only in its appointed place, but also arranged with millimetric accuracy. If we should ever buy another Morgan, we would hope that Brett will build the rolling chassis.

Chassis erection and finishing shop in Pickersleigh Road, 1919.

Chassis Shop

Brett is responsible for building the rolling chassis of our Morgan. For the past fifteen years, he has worked as a mechanical fitter in the chassis shop. The clock on the shop wall shows 11:20, and he has been working on the car since 10:10 – the end of the morning tea break.

The allotted time for creating the rolling chassis is about 515 minutes, in other words about eight and a half hours. That, however, is the only stipulation. The order in which he assembles the individual parts is up to him. There are few other directives he needs to follow; every worker is independently responsible for his own area of specialisation. The basic element, called the z-frame-chassis, is sourced from a supplier in Ross-on-Wye, 35 miles away. The chassis, of galvanised steel, are neatly stacked outside, in front of the shop. At the beginning of assembly, each chassis is rolled into the shop on dollies and, with the help of a crane, lowered onto a pair of trestles at the work station.

As a preliminary operation, Brett has already attached a couple of elements, taken two wooden floorboards from a stack against the wall, and installed clutch and brake pedals. He has mounted the suspension bar, a black metal hoop to which other parts will be attached, and installed sections of the brake pipes as well as the handbrake lever.

MAKING A MORGAN · DAY 1

The individual chassis kit for order no. 92183 is now complete.

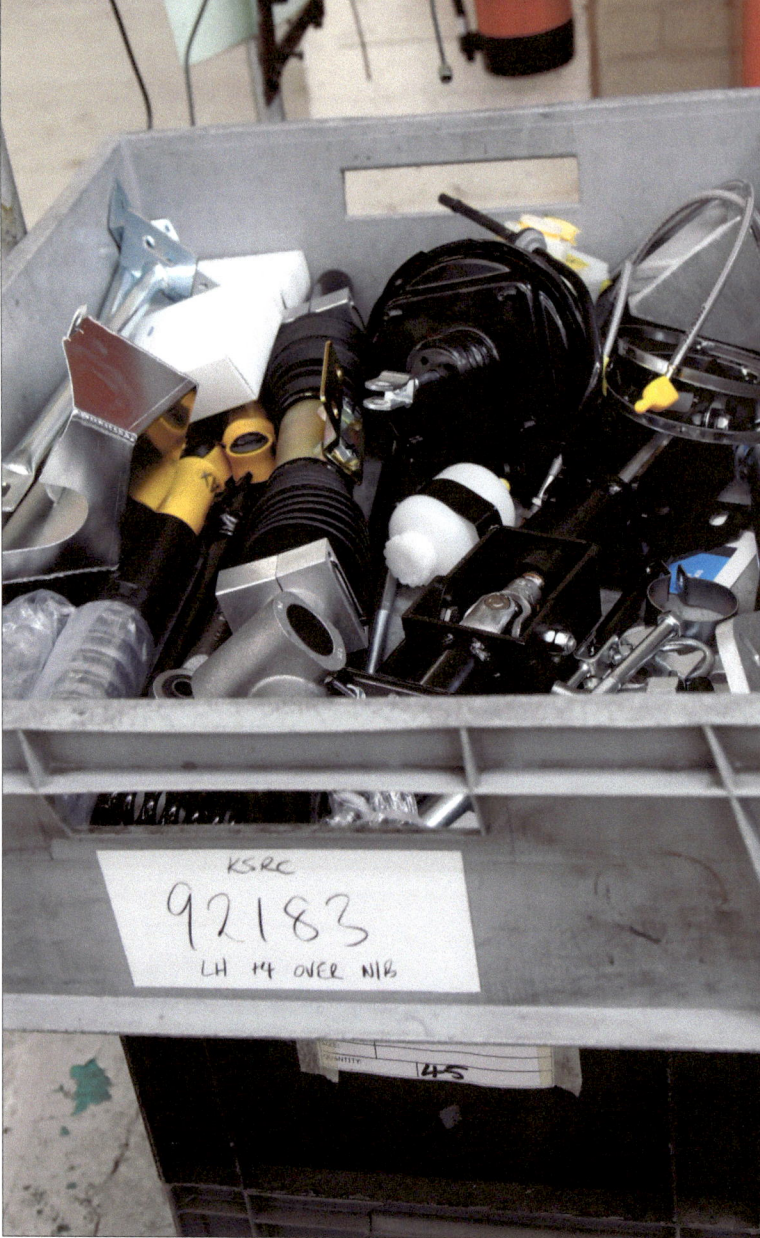

Apart from these elements, the chassis still looks quite virginal. Lucky for us, since we don't want to miss an opportunity to lend our own hands to the manufacturing process. First, Brett installs yet another wooden floorboard.

All parts that represent the as-ordered specifications and not stored as standard components around Brett's work station are supplied in the chassis kit, a grey box delivered by Sam and Jamie from the goods in – goods out shop, along with their routing papers.

Machine Shop

The machine shop, under foreman Mark Cerrone, is one of the departments that don't get their fair share of credit because they deliver parts to all of the workshops. In the machine shop, wheel hubs, hood hardware, and brake drums, among other items, are either fabricated, modified, or assembled from parts provided by outside suppliers. The staff in this hall consists of nine people.

In a corner of the sheet metal shop, assembly workers sometimes weld their own elements. The machine shop includes a number of wonderful old machine tools, such as pre-war lathes that still carry out their tasks, reliably and without complaint. But peek around the corner and you will hardly believe your eyes. What would have been unthinkable during our previous visit, four years ago, has become reality: modern CNC cutting and milling machines!

We faced a real dilemma – should we photograph these, and so destroy a small piece of the Morgan mystique? But we are of the opinion that – along with the venerable old shop halls as well as the ever-present artisanry of traditional craftsmanship – die-hard Morgan fans should tolerate these few concessions to modern manufacturing conditions. After all, they do contribute to economical workshop operation, and therefore ensure the future of the Morgan marque. And based on those considerations, we will make minimal revisions to our statements regarding the topic of totally old-fashioned manufacturing in our books, blog, and presentations.

How it all begins: The chassis rests on trestles as it is fitted with the first of many parts. The black suspension bar is seen at the right.

Someday, this will be 23 new Morgans. Galvanised chassis stored outside the factory halls.

Driveshafts and drivebelts dominate this view of the Worcester Road machine shop, 1912.

Ancient but in top condition: This lathe has seen decades of service.

Protected by purple plastic: brand-new front hubs.

The machine shop also has representatives of the latest generation of machine tools.

They don't make them like this anymore: a veteran Milwaukee horizontal milling machine.

It was not so long ago that CNC machines would have been unthinkable at Morgan.

Test-fitting disc brakes in the machine shop.

A spirit level is used to check the alignment of parts.

With a practised eye and a great deal of experience, holes are drilled in the narrow frame front.

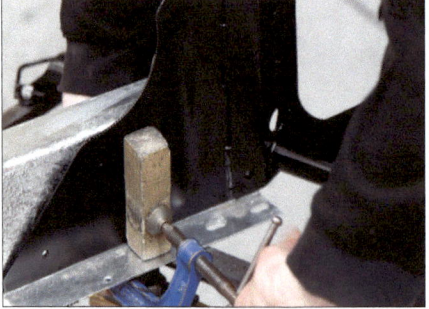

C-clamps hold the front in place during assembly.

The brake master cylinder is installed.

Chassis Shop II

The first important component to be mounted on our chassis is the aforementioned narrow frame front. Later, the steering and front suspension will be attached to this. Dimensionally precise steel bars and a spirit level ensure millimetric accuracy in fitting this component.

Prior to final bolting, it is held in place by screw clamps. Next, Brett installs reinforcements – so-called straining bars – on each side.

As many elements or patterns (jigs) have been specially made they carry unique designations in the plant. For simplicity, instead of the official designation we will explain the function of the part along with its designation.

Returning to the assembly process, the next step is to mount the clutch and brake master cylinders to the chassis, followed by additional brake pipe components. To bend them into the correct shape, Brett uses a prepared

Brake pipes are bent to shape using templates.

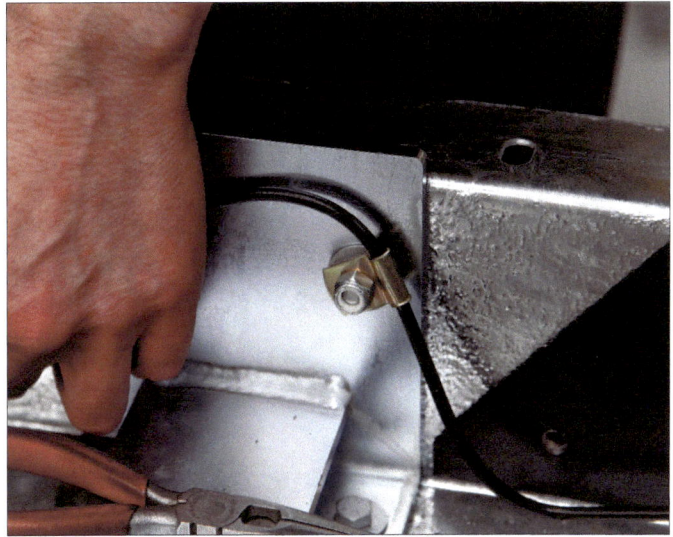

The brake pipe is attached to the chassis.

Repeated fitting and checking is part of the assembly process. A few gentle taps of the hammer help to seat the steering rack.

template made of sturdy iron wire. By hand, and with a practised eye, he bends the brake pipe components to match the template and attaches them to the chassis. Next, the steering rack, along with the steering dampers, is attached to the front of the chassis.

The workers return to their workplaces with the same punctuality with which they left. After lunch, Brett continues to work on the brake pipes, and attaches the

Brake hydraulic line three-way connector.

Your reporter is allowed to lend a hand. Carefully, the engine and transmission are hoisted …

… and then gently lowered into the chassis.

The rubber engine mounts isolate the Ford four-cylinder engine from the chassis.

supply line to the clutch master cylinder. Next, and to my complete surprise, I add my contribution to the rolling chassis. Brett asks me if I had ever installed an engine. I admitted I had not, at which point Brett suggested that today was a good day to make the attempt. What appears to be a very complex operation is in fact relatively simple. Brett attached the selected engine to a transport crane by means of chains. My task consists only of hoisting the engine and manoeuvring it over the chassis.

Then it's a matter of maintaining the correct installed position, and that's precision work, requiring millimetric accuracy. First, two engine bolts, in the transmission area, need to be guided into drilled holes. Carefully wiggle them into place … done! Fortunately, the drilled holes are correctly located. Now, all attention turns to the bolts below the engine mounts. There, too, the location is correct – even though it takes a little bit of persuading. Now lower the engine all the way … it fits! Finishing off, four bolts are installed, and the job is done.

After my little sojourn into the 'real' working world, I return to the familiar tools of my own trade – camera and pen. And Brett turns to the clutch, connecting the clutch hydraulic line.

Next he addresses the front suspension coil springs and the front disc brakes. He arranges all the parts from the

Front suspension springs.

The washer won't fit … It needs to be bent. Not a problem.

appropriate box on the floorboard. The parts are fitted, lubricated as necessary, and installed.

Now and again, individual elements such as washers don't fit as precisely as desired. Some improvisation is called for – the washer is clamped in a vice and bent to fit. Anything that doesn't fit, will simply be made to fit. We'll encounter this principle time and again as assembly progresses.

Break time

At precisely 12:25, a decidedly old-fashioned handbell announces the midday break. In the blink of an eye, the hall empties. The break lasts until 1 o'clock. At Morgan, breaks and work periods are strictly observed, each announced by the same handbell signal. Work begins at 8am, ten-minute tea break at 10am, 35-minute lunch break until 1pm, at 3:25pm another ten-minute tea time, and, precisely at 4:30pm, it's quitting time. Tidying up of work stations doesn't begin until five minutes before quitting time. On Fridays, the workday ends at 1:30pm, and instead of a ten-minute morning tea and lunch break, there is a somewhat longer 15-minute tea break at 10:45am.

MAKING A MORGAN · DAY 1

The bulkhead (firewall) is installed – and suddenly our Morgan stands taller.

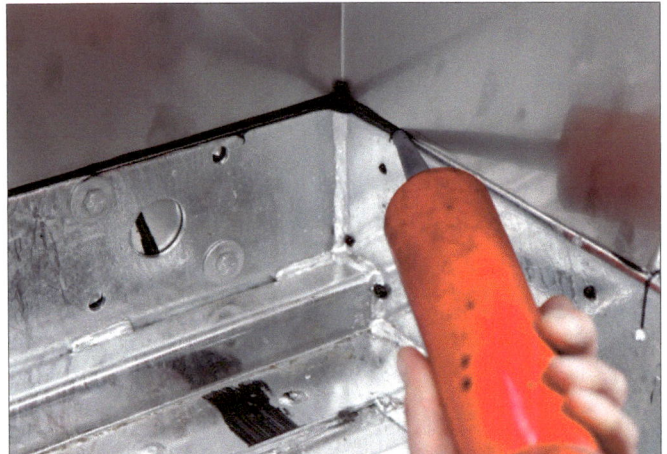

Industrial adhesive bonds tightly and also seals the footwells.

Rivets are used to join the bulkhead sections.

First fill: the stub axle is greased.

Completed assemblies are colour marked to avoid repeated re-checking.

Front suspension: spring, Spax shock absorber, disc brake and centre-locking hub.

Independent front suspension

Morgan's present-day independent suspension is still based on H.F.S.'s designs of 1910, when he was building his first vehicle. The original three-wheeler, too, had independent suspension, which remained largely unchanged throughout the production life of the model, to its termination in 1952.

From the outset, the first four-wheeled Morgan of 1936 got this same type of front suspension (with minor modifications). It remained even through the 1954 facelift that gave the 4/4 its present-day form, with its integrated headlamps. Of course, there were additional minor modifications, such as modern Spax shock absorbers.

After installation of the shock absorbers, the first of several grease nipples are addressed. To our knowledge, the Morgan is the only car being made today that still uses grease nipples that will demand regular lubrication, every one to two thousand miles, depending on operating conditions.

Before going any further, it's time to see red – all major components whose installation has been completed, are marked with red paint, so that if necessary, any other craftsman in the department can carry on with assembly without needing to check all of the existing bolts and threaded connections.

MAKING A MORGAN · DAY 1

Quitting time. If this were a cookbook, it would tell us to let the dish sit overnight. After the first day of work, the chassis is already beginning to look like an automobile.

The exhaust manifold is installed with a few twists of the ratchet.

Things are a bit too tight here …

… but the tin snips will soon see to that.

Now Brett turns his attention to the bulkhead. This consists of four stainless steel sections that are positioned and then riveted together. Industrial adhesive is liberally applied to all joints; this serves a dual role as a sealant. When driving in the rain, the finished Morgan's occupants are entitled to dry feet: water should be excluded from the interior. From personal experience, we can confirm that this works. In heavy rain, there will be some leakage at other locations, but that's a different topic. After all, we are driving a hard-core roadster, and not a plush, luxurious convertible. The finished bulkhead is first clamped in place and then bolted to the chassis.

The next component is the exhaust manifold. On the four-cylinder, this joins a single-pipe exhaust system. Before installation, a single heat shield needs to be installed. The manifold itself is done in a matter of minutes.

After the afternoon tea, the chassis takes another step toward a recognisable automobile – aluminium side panels, so-called valances, are installed on each side of the engine. Because the clearance to the exhaust system isn't quite right, Brett simply uses his tin snips to cut away a piece of aluminium on the passenger side. Whatever doesn't fit, is made to fit – but we said that already.

The last action before quitting time is mounting the cowl box, which echoes the sweep of the classic radiator grille, which it will later carry. That's all for today. The results are respectable: a nearly complete rolling chassis.

Day 2

Part of the outsourced steering column is in the way, but is quickly addressed with a cutting disc.

By morning of the second day, the grey parts box is nearly empty.

The chassis is put on wheels

Naturally, we didn't want to miss a thing, and so we appeared at our station just before eight o'clock – just in time to see Brett connect more brake pipes. Even at this early point in the assembly process, a glance into the box of parts to be installed reveals plenty of empty space.

Now the time has come to connect the clutch and brake pedals to their respective systems. The largest part remaining in the box is the steering column. To attach this, corresponding mounts need to be affixed to the bulkhead.

But the steering column doesn't fit perfectly. Part of the housing is hampering its installation, and Brett cuts it away with a disc grinder. The supplier has yet to deliver a steering column that fits, so the installer has to make it fit. To install the steering column, Brett uses one of the many holding fixtures. At its other end, the steering column is joined to the steering rack, which also carries the steering damper.

In the meantime, Nigel has finished the pre-assembly of the complete rear axle, with brake drums and leaf springs. He uses a crane to place the axle on the chassis. The axle is bolted in place, and the leaf springs are attached to their shackles. To do this, the springs need to be pre-tensioned by means of a long lever. The shock absorbers as well as the battery tray are attached to the now-familiar suspension bar.

MAKING A MORGAN · DAY 2

Nigel mounts the drum brakes to the rear axle.

The rear shock absorbers are bolted to the tubular crossmember.

The finished rear axle, with brakes and leaf springs, is lowered onto the chassis.

The leaf spring is attached to the axle. The excess lengths of the U-bolts are also attacked with a cutting disc.

Now Brett goes away for a while; he has to get a battery from the warehouse. After all the recent activity, this seems to take forever. The handbrake cable is also attached to the suspension bar. A splitter to distribute hydraulic pressure to the rear drum brakes is attached to the live rear axle housing.

By the way, these drum brakes are not just stubborn adherence to old habits, but rather they work amazingly well. We tested this for ourselves, including driving at the limit on the Le Mans race track.

Brett fills the differential with gear oil. One vital driveline part is still missing – the propshaft. This can

The battery is located ahead of the rear axle. Later, it will be much more difficult to access this area.

The rear brake hydraulic splitter is attached to the axle.

Propshafts in a violet-coloured steel mesh container – called a 'stillage' – await installation.

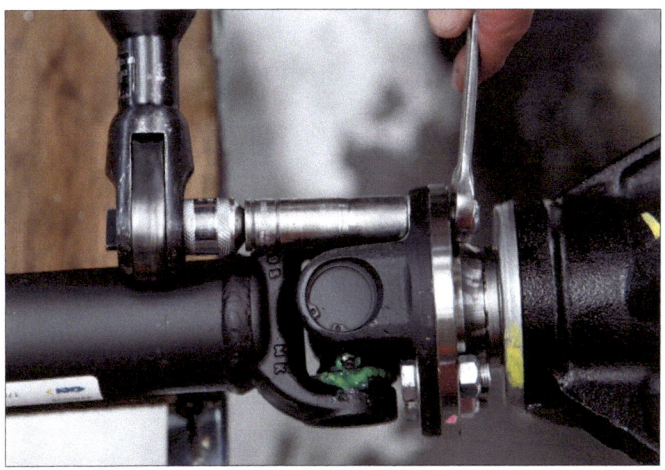
The short propshaft is bolted directly to the differential flange without any additional damping.

be found in a 'stillage' – a violet mesh box for larger components, alongside the work station. In seconds, it's placed in position.

Next, Brett would like to install the steering linkage, but before he can do that, the familiar tea-time bell sounds. As we have learned, the workmen hold their breaks sacred, and instantly set their work aside at the sound of the bell.

All right, so the linkage can wait another ten minutes. Once its four mounting bolts are tightened, the steering is fully functional.

Looking in our chassis kit, we see there are still two cartons from supplier Ford, with the two oxygen sensors to be mounted on the exhaust system. There, all the parts are installed. That will do it, for now.

To make this truly a rolling chassis, it still needs wheels. But first, brake fluid needs to be added. It takes a while for it to flow through the long, tiny brake pipes, and the first drops appear at the bleed nipples. A colleague helps Brett by working the brake pedal to build up pressure, until the system is completely filled.

Meanwhile, the foreman of the chassis shop, John Burbidge, takes care of mounting and balancing the tyres on the

MAKING A MORGAN · DAY 2

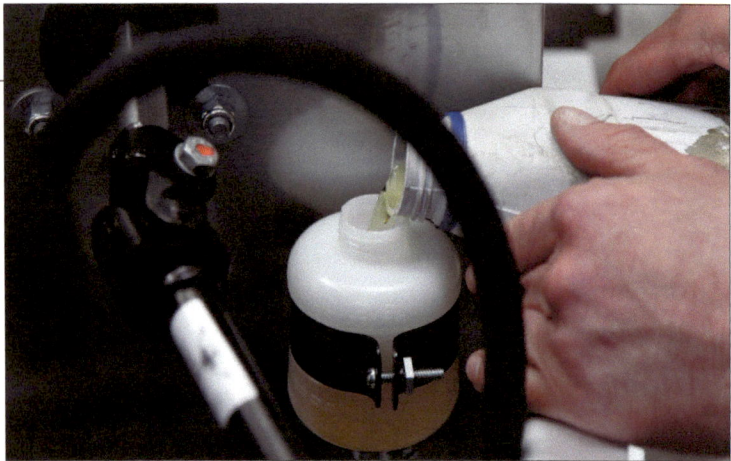

The brake system is filled with fluid. In two years, it will have to be replaced as part of routine maintenance.

Once the two oxygen sensors are installed on the exhaust system, the grey box is empty.

Foreman John mounts the tyres on gleaming wire-spoke wheels.

sparkling chrome wire wheels. Here, the boss still lends a hand. The Avon High Speed tyres, size 185/70 R 15, seem downright dainty and undersized by modern standards, but are a perfect match to the appearance of the traditional Morgan.

Although 'our' rolling chassis is now complete, it still needs to pass function checks. The handbrake is carefully adjusted. Brett pays special attention to setting the front toe-in. Over and over, he checks and adjusts, until finally he is satisfied.

The clock reads 11:30 as he uses the hoist to lift the car from the bucks and pushes it, ready for pickup, to the ramp leading to the assembly shop. The build book records what has been done so far, and John does one last visual check, while Brett has already turned his attention to the next chassis. The parts box for that job is already in place. This time, it will be a six-cylinder roadster.

The clock reads 2:10 as Karl lets the car roll down the ramp to the next work station. Actually, he glides down the ramp, more like a skateboarder, but backwards, and stops the car with the handbrake. He has to push it for the last few feet. Karl uses a floor jack to lift the car, and now we eagerly await such exciting events as the impending

To continue the assembly process, the chassis once again has to be hoisted onto trestles.

Now, it's officially a rolling chassis. Karl rolls it backwards down the slight gradient into the assembly shop.

The fuel tank is quickly installed at the very rear.

'marriage' of body and chassis – truly, a bond for life – the life of the car, at least. With his own unique dynamism, Karl quickly sets to work and in the blink of an eye, has installed the fuel tank.

Then comes the news that there is no completed narrow body, as specified for our car, in the storage rack. In truth, the aluminium-covered wood frame should have been completed long ago – especially since 'our' Morgan had been given top priority in the production order. But there is no appropriate body to be seen. So the rolling chassis will have to mark time in the assembly shop.

The upside of all this is that we will now have an opportunity to follow the creation of a body in the wood shop and in the sheet metal shop – and in fact it will be 'our' body, and not some random warehouse part. Let's set the topic of the assembly shop aside for now, and turn our attention to the body, to its completion by the supplying departments.

MAKING A MORGAN · DAY 2

Nic has coated the layers of wood for the inner rear wing with glue, and clamped them in the bending form.

The left rear inner wing takes shape.

Traditional craftsmanship: Vince working on the inner rear wings with a mallet and chisel.

Wood Shop

At 3 o'clock we arrive in the wood shop, where Vince has already begun assembly of the 96 shaped wooden parts that, even today, make up the legendary ash frame of a traditional two-seat Morgan. The job will take 13 hours; at the end of which, the wooden frame, with its personal identifying number of 552, will be completed.

The wood shop is the workplace of eleven people, in three work areas. The first area, with four woodworkers, is called the mill. There, the pre-delivered wood is sawn to size and individual parts undergo preliminary shaping. It is not only for safety reasons that this shop is off limits to the countless visitors; here, sight of the most modern woodworking machinery might chip away a bit at the 'Morgan mystique.'

Vince Wanklin – bedrock of the wood shop

At age 64, Vince is one of the 'bedrocks' of Morgan. He, too, is a native of Malvern. After working as a cabinetmaker, he found himself out of work – only temporarily, as it turned out. A conversation with Peter Morgan led to a job in the wood shop on the very next day, assembling wooden frames.

That was almost exactly 40 years ago. An archival photo, happily displayed to visitors, shows him with a head of hair that would do Jimi Hendrix proud. In a few months, he will take his well-earned retirement. He loves the peace and quiet, and the cleanliness, of the wood shop and would never trade jobs with his colleagues in the metalworking departments.

Perfect wooden frames are his world, in the past and today. He tends to his work with appropriate care and great attention to detail. In his free time, he is first and foremost a family man. He and his wife have three children and five grandchildren. Added to these are eight nieces and five nephews – to say nothing of the more distantly removed members of his extended family. He has never driven a Morgan.

Vince enjoys his job; he's been doing it for some time. This picture was taken 40 years ago ....

Next, the finished elements are either placed in special containers near the work spaces, from which workers will draw them in the course of assembly, or in that corner of the space where Nic builds up entire assemblies from individual parts. Another work station is for making special dashboards, with exotic wood surfaces. The remaining six workers assemble the wooden frames. All are qualified to assemble two-seaters, but the four-seater and Aero series are reserved for specialists.

The wood shop is definitely a Morgan institution, and one of the important differences compared to other automakers. The shop exudes a certain calm. There are no loud machinery noises, apart from an old radio that plays

In the wood shop, 96 different wooden parts are joined to make the frame. Here, Vince uses a template to drill holes in the inner wheel arch closing panel.

MAKING A MORGAN · DAY 2

Just a few more holes to drill …   … and the inner wings are ready for mounting to the frame.

Vince is satisfied only when all dimensions and angles are correct.

nothing but oldies, and is only turned off during breaks. Sometimes, one of the workers will sing along – or, like Vince, whistle along – to a familiar tune.

Vince is assembling the parts for the rear wings. He uses the laminated wooden members made by Nic using the now half-century-old bending forms. To make such a part, Nic applies ample amounts of carpenter's glue to three boards, and forces them into the bending form, which is then clamped firmly. They stay there for at least two hours, until they have permanently taken on the required radius. Because their trailing edges taper to a point, Vince saws them to shape – using an experienced eye.

The two rear wing components are joined by a transverse member, the so-called hood release. This step defines which hood will be installed on our car; in our case, it is the modern hood, whose rear can be unlatched for easier closing. A surprising side effect is that die-hard fans of vintage cars are often surprised to see how easily the generously dimensioned luggage space behind the seats can be accessed and loaded. Then, at exactly 4:30pm, the now-familiar bell announces quitting time.

Day 3

Precision in wood: stacks of finished elbows and other parts.

'The elbow' – or: small cause, big effect

The next morning, after just 20 minutes, work suddenly screeches to a halt. A common, pre-formed component, which should be installed at this point, can't be found in its customary space on the shelf.

Actually, the elbow is just one component of many. The fact that it gets special treatment here is only due to a defective drive belt on a modern woodworking machine in the mill. Because of the Monday bank holiday, a replacement could not be obtained in time to avoid a halt in production of wooden shapes. Production continued as usual – only without the aforementioned elbow. The machine was repaired quickly enough, but sufficient numbers of the needed parts were not on hand until after the noon break. Of course, Vince and his colleagues couldn't just mark time, waiting. The wood shop had enough half-completed parts and frames on hand to keep everyone occupied.

The only one showing any impatience today is the author – that was until I had extracted a promise from the workers in the mill that they would have the missing parts done by

All in order again: Production of the elbow can proceed.

1:45pm. To the cheers of the crew, I was personally allowed to roll the cart with the keenly sought elbows into the wood shop, and turn them over to Nic, who went over to the preparation area and assembled them, post haste.

Of course I kept a close eye on his actions as he clamped the individual parts in a jig and glued them together. I immediately grabbed the first two examples, for the driver and passenger side, and carried them over to Vince, who turned his attention back to 'our' frame. The elbow is the connection between the rear wings and the doors, which could not be installed without it.

Nic coats the surfaces to be joined with white glue.

The individual parts are clamped into position …

… and joined by glue and screws.

Meanwhile, Vince has been working on the doors and the impact bars – sturdy metal components to which the side impact protection will be mounted later. Then – finally – he adds the elbows. They are still missing two tiny wood pieces, but Vince happens to have a few of these in the secret stash under his workbench.

Dimensional accuracy of the elbow is vitally important, as the doors are very sensitive in this regard. The standard for accuracy is quite high, but after a few small corrections, everything fits perfectly – as a critical glance at the gauging tools quickly confirms.

Now a transverse member on the passenger side, to which the latches for the hood will later be attached, a few wood and metal reinforcements, a few cut-outs, a few drilled holes, time and again dimensional checks – and the frame is (nearly) finished.

Nearly, because the next step is fine tuning. The smallest imperfections are attacked with chisels, planes, and sandpaper, until everything is as smooth as the proverbial baby's bottom. Vince gives the underside of the frame the same attention.

Ultimately, nothing of the frame will be visible. But Vince frets over the wooden core as if it were a piece of furniture for Buckingham Palace. Something important is still missing – the doors! But once again, the time has come to sweep up the day's wood shavings. Quitting time in five minutes.

'The elbow'

Meanwhile, Vince continues working on the door frames.

And again: the practised eye and an incredible amount of experience guide the hand of the craftsman.

The template fits, all is correct.

No, children, this is not a computer-controlled electric tool, but rather a hand plane.

A router guided by a template is used to create recesses …

… such as this mortice, where a part will be installed later.

MAKING A MORGAN · DAY 3

A template ensures the accuracy of …

… a few drilled holes.

For added grip, glue is applied to threaded inserts …

… before they are installed in the frame.

A digression: From ordinary plywood to exotic wooden dashboard

The dashboard of 'our' Morgan will have only a painted surface, but we would also like to examine a very special process that magically transforms ordinary plywood and a thin layer of veneer into a small work of art. Time and again, beauty of its wood grain will enchant all who gaze upon it.

It all begins with a piece of plywood from the well-stocked wood warehouse. The mill cuts these into convenient rough shapes, and returns these to storage. Whether any given blank will become an exotic wood dashboard or a standard production one, is yet to be determined.

Steve looks at the work orders pinned to the corkboard for that day's orders. On this Tuesday, dashboards for left- and right-hand-drive cars are called for. These will require a number of jigs. Steve guides the machine head along the groove in the jig, as the machine cuts the contours – including cut-outs, such as those for the handgrips.

Next, the edges are smoothed. The lids for the glove boxes are sawn out separately, and worked the same way. For most dashboards, the cut-outs for the glove box and instruments are done in the earliest stages. For later veneer work, without cut-outs, blanks are pre-cut and stored until needed. In the wood shop, David selects the appropriate veneers and arranges them so that the grain at the central joint presents an attractive overall appearance – if possible, as mirror images (so-called 'book matching').

A single piece of veneer is seldom sufficient to cover an entire dashboard. David joins the selected sheets with special adhesive tape. Then the veneer is turned over, the dashboard placed on it, and weights are applied. Using a sharp knife, David cuts the veneer to shape, allowing a bit of excess. The blank is coated with glue and the veneer carefully laid on top.

Just as carefully, David lays the dashboard, with the visible side up, in a plastic bag, to which vacuum is applied. By means of this amazingly simple technique, the excess

The well-supplied wood warehouse contains birch plywood in several different thicknesses.

The upper shelves hold Plus 8 dashboards for left-hand-drive US versions.

A router jig is used to shape a dashboard blank.

Getting fancy: David cuts the wood veneer to size.

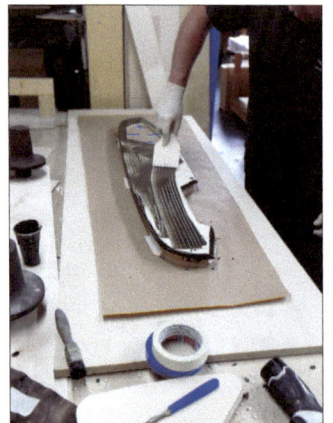
Glue holds the veneer layer to the dashboard – which, in Morgan's case, is truly a wooden board.

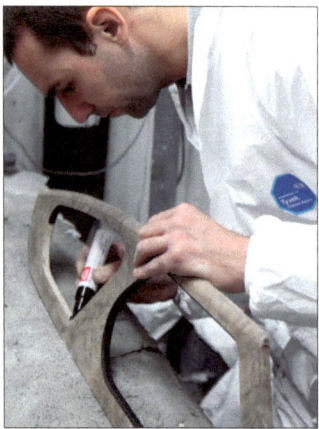
The visible edges of cut outs and the back of the glove box door are coloured by means of a special felt-tip marker.

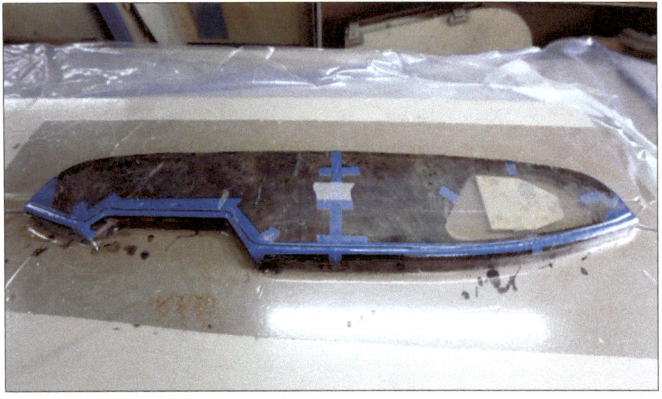
Vacuum bagging presses a plastic film against the veneer and moulds it to the board as the glue cures.

Steve neatly routs out the holes required in the dashboard.

material of the veneer flawlessly wraps over the edges. The curing process, aided by infrared lamps, takes several hours.

Now Steve in the mill takes over again. He routs out the holes for the instruments, as well as the glovebox, and works on the sharp edges. Then it's once again the turn of David in the wood shop. He primes all the edges and mounts the hinges as well as the glovebox door. The dashboard is now essentially done, but doesn't look especially spectacular. That will change at the next work station.

In the paint shop, Mark once again disassembles the glovebox door and marks all individual parts. Then he colours the primed edges, as well as the backside of the glovebox door, using a special, extra wide, black felt tip marker – a method that is as simple as it is effective in preventing thinned black paint from penetrating the veneer surfaces and so ruining the workpiece.

Mark screws a temporary grip to the board, which will allow him to hold and turn it without damage while applying finish. Clear lacquer for the first layer is mixed to an old secret recipe and applied as a heavy layer in the tiny spray booth.

After drying, this filler coat is sanded down, in five stages, by hand and by machine, to an exceedingly thin layer. The sanding grit is successively finer, from 20 grit all the way down to 1000.

A digression: From ordinary plywood to exotic wooden dashboard

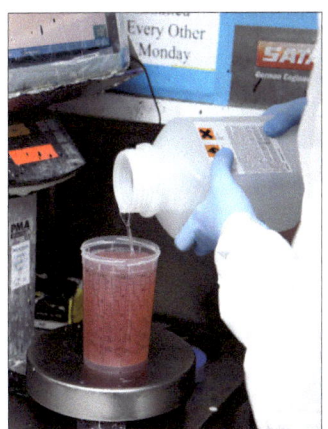
The exact recipe for the first layer of lacquer is a secret.

A shining example: spraying the first, heavy coat of lacquer.

The secret of a genuine, deep shine: keep polishing.

A microfibre cloth sweeps away the last blemishes.

Finished. Mark is satisfied with the new showpiece …

Next, Mark seals any tiny pores, hardly visible to the unaided eye, with filler, and wet polishes these spots with multiple applications of special polishing pastes. Then the edges are once again treated to the oversized black felt tip marker; after the final application of clear lacquer, this will rival the finish on a grand piano.

Finally, lacquer for the cover coat is mixed and applied as a very thin layer in the spray booth. In the process, a fly somehow got into the booth, unnoticed at first. It resisted all our efforts at capture. We could only hope that it would not land on the freshly lacquered dashboard, or on a wing, as these were drying. We were lucky; upon opening the doors, the fly could not be found. Otherwise, the affected piece would have to be prepared all over again, and refinished.

Another go with the special polishing paste – with grits as small as 4500. After reassembly of the previously marked hardware, Steve lovingly polishes the minor masterpiece one more time with a microfibre cloth. Then it goes off to final finish, where Jo is waiting, a little impatiently. Unfortunately, this dashboard is not intended for 'our' Morgan.

Day 4

Work continues in the Wood Shop. Vince test-fits the door frames …

… and continues to correct their shape until everything fits perfectly.

The wooden frame is finished

At 8 o'clock, right on time as always, Vince begins his workday by installing the doors. What may sound like a simple operation is actually a complex process requiring submillimetric accuracy. On our particular Morgan, the doors are unfortunately a very critical element; ours had to be re-fitted several times.

The fundamentals must be right. So Vince repeatedly re-fits each door, reworks this or that spot with the plane, sands, checks again and again with his hands, takes off a tiny bit more … until the door finally has the optimum fit.

Then he uses a chisel to make cut-outs. The chisel has to be re-sharpened several times during the delicate process. Finally the hinges are attached to a rigid iron plate on the frame. To do this, Vince drills the frame at pre-marked locations. He also drills and threads holes in the iron plate.

For most holes, a machine with an integral tap is used. One hole, however, has a different diameter, so Vince has to use an old hand tap – totally old school, but not at all uncommon at Morgan.

The wooden frame is finished

Pillar drill creating screw holes in the brackets

The hinges are mounted.

This plywood sheet, with the door cut out, helps the Tin Shop crew in cutting the door skins to size.

Assembly line à la Morgan. The frame is hand-carried to the 'bathroom' to be dipped in preservative.

After nearly three hours, the doors have been installed with precision, and in a final step, the frame number – 5552 – is stamped in two places. Then a special template has to be prepared for the tin shop. Because each door is unique, individually fitted to the frame, the template serves to cut the matching aluminium door skin.

After a bit more paperwork in the accompanying papers, Sean helps Vince push the frame across the yard to the so-called bathroom, where it is completely immersed in a vat of preservative called Protim 800. In the past, Morgan frames did have a tendency to rot, and had to undergo complex restoration. That is now a thing of the past. The frame cannot be left in the vat for too long; after just three minutes, it is pulled out and allowed to drip over the vat.

That was it for Vince's contribution. At 11:30, the wood frame, number 5552, is history as far as he is concerned. He is one completed frame closer to retirement, and he's finally rid of that pesky writer fellow.

MAKING A MORGAN · DAY 4

The preservative is called Protim 800. It helps make the wooden frame much more durable.

Three minutes in the bath ensure a long service life. But first, excess preservative needs to drip off the frame.

The Sheet Metal Shop, 1919.

Just a word with Graham Chapman on 'quality'

Graham has been technical director at Morgan for the past ten years. His background is in the manufacturing industry, and when it comes to manufacturing quality at Morgan, he carries primary responsibility. When he speaks to veteran Morgan drivers, he hears their complaints: the wood frame rots, the hood leaks, or the electrics are unreliable. Almost always, the cars in question are older specimens. Even in our own personal Morgan, built 2011, heavy rain will result in a few drips into the interior. But we don't obsess about that; after all, we are not driving a mass-produced sedan built by robots, but rather a hand-built, hard-core British roadster. We don't expect door gaps like those Ferdinand Piëch once demanded of the VW Phaeton.

Still, quality control is an important topic, one that occupies Graham and his crew from morning until evening. How can one reconcile the quality demands of our fully automated age with traditional craftsmanship à la Morgan? When Graham came to Morgan, the equipment was seriously antiquated. Since then, the company has invested heavily, for example in CNC machines in the machine shop, a computer-controlled press brake in the tin shop, a 3D printer and scanner for development and quality improvement. Graham's primary goal is to make all elements of the process perfectly repeatable – without reducing the share of hand craftsmanship. To that end, a small 'development department' was created, mainly tasked with detail solutions. Its task is to work on improvements to the current models, changes for which the research and development department no longer has excess capacity.

Simultaneously, all processes were precisely defined, every component and its dimensions analysed three-dimensionally by computer, and, wherever possible, new jigs were created to further optimise the steps in the production process and make them as repeatable and free of errors as possible. This includes, for example, placement of pre-drilled holes. Jigs – by now numbering in the hundreds – have long been used in Morgan's shops, but these had been made on the initiative of individual workers, and not systematically or deliberately as part of production planning. One worker from the manufacturing department was trained as a quality assurance engineer, and put to work – a first in the history of the Morgan factory. At the end of every production process, there is a 160-point quality check, intended to guarantee the quality of work.

At Morgan, as part of the production process, there is a sort of customer/supplier relationship between the various shops. Production is a co-operative process, and every shop depends on other shops supplying fault-free components and assemblies. Today, if that is not the case, the offending part is immediately sent back for rework. Overall the depth of manufacturing has been reduced, and production of parts that can be made more precisely on modern machinery, outsourced. One example is pre-formed wings, now supplied by Superform Aluminium of Worcestershire. One vital contribution to improved quality is the fact that today, the workers can more easily implement their own ideas for improvements. Ultimately, they, as a whole, represent a knowledge base of several thousand years. In the past few years, the sum of all these measures has led to a readily perceptible improvement in detail quality.

Also, there won't be such hasty actions like the one in 2011, when the new 3 Wheeler was shown at the Geneva Auto Show before production had even started. At the time, the sudden influx of 300 orders caused the first cars to be delivered before the model had been properly developed and thoroughly tested. This in turn led to considerable quality flaws, and enduring frustration on the part of new owners. For that reason, on the new Aero 8, Morgan is taking the time to conduct comprehensive testing and continued detail development, before the first examples are delivered to their new owners.

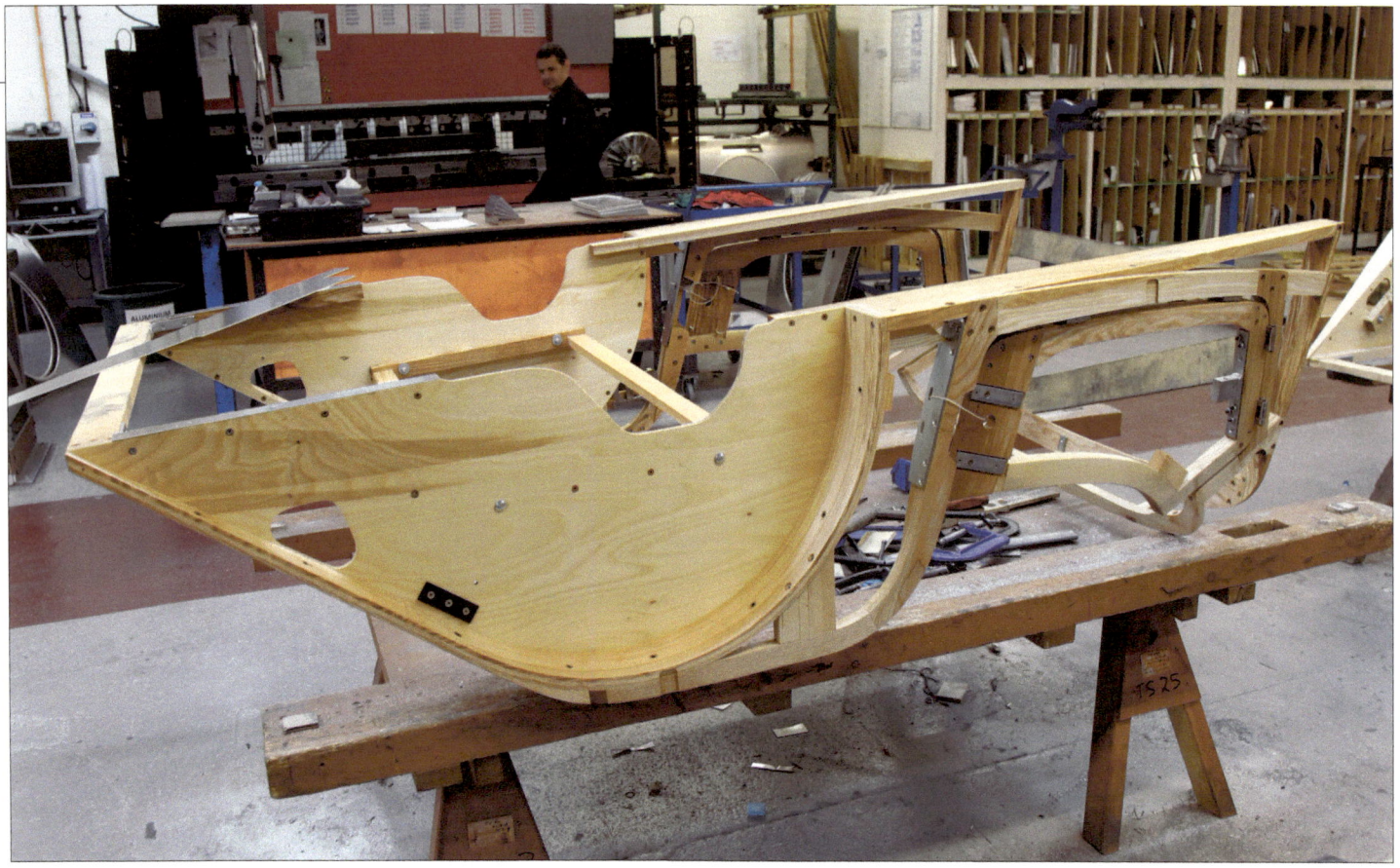

The wooden frame stands ready in the Tin Shop, where it will be fitted with its metal skin.

Sheet Metal Shop (aka the Tin Shop)

Preliminary word indicated that the frame would not undergo any further work until the next morning, but at 3:30pm, we were told that the build process would continue in the tin shop. Richard already has the frame up on the stands, and begins the task of skinning it in aluminium.

On first seeing this, I experience a sort of culture shock. After witnessing the love and care invested by Vince as he built the doors, I can't believe my eyes. Using the template made by the wood shop, aluminium sheet has been cut in the shape of the door frame. The rear is simply brushed with anticorrosion paint, and then just roughly nailed, tacked, and screwed to the oh so recently immaculate door frame!

How on earth are we going to get a perfect surface out of this? Not to worry; firstly, this sheet will be carefully sanded, after the doors are removed by knocking out the hinge pins, and secondly, this is only a substructure. The actual door skin will be fitted with great care and a great deal of hand craftsmanship.

In the process, the edges are worked using an old, hand-cranked bead roller. Then begins the painstaking process of using a hammer to hand-work all the sides, and above all the curves. And once again, it's quitting time.

Sheet Metal Shop (aka the Tin Shop)

Richard goes to work …

He tacks, nails, and screws …

… and repeatedly sands down the high spots.

Using an abundance of care, the actual door skin is installed over the inner skin.

This bead rolling machine has seen many years of service.

The curves in the door skin flanges are worked with a body hammer and hand anvils – 'dollies.'

Hammers, files, and dollies – the tools of the panelbeater's trade.

Sheet Metal Shop – second day

Richard again turns his attention to the doors. Now the aluminium skin begins to take shape. Skilfully, Richard works the metal, using an amazing array of hammers. As hectic as Richard appeared while attaching the door substrate to the frame, his work now is much more deliberate, relaxed, masterful.

After the obligatory anti-corrosion treatment, he uses six hammers and two files to roll and crimp the especially narrow edges. With every blow, he changes the angle of the hammer ever so slightly. After peeling off the protective plastic membrane from the aluminium skin, the edges are filed and sanded. Wherever attachment by means of crimping is not possible, he uses very small nails. This is followed by the final finishing by means of careful filing and sanding.

At the end, a test – does the door fit the frame? It does. At 9:30am, both doors are done. Time required: 2 hours, 45 minutes. In all, fifteen hours in the tin shop are allowed to make each body. That is a tight schedule, but the Tin Gods have their pride and would rather earn a rebuke from Conrad, head of the department, than deliver defective or even incomplete work as a result of time pressure.

Sheet Metal Shop – second day

It takes years of experience to shape door skins with the ease displayed by Richard Harris.

Richard Harris – the 'Tin God'

Richard is 26 years old and, of course, also a Malvern native. At the age of 16, right after completing his final year of school, he entered into a five-year apprenticeship at Morgan as a sheet metal worker, also known as a 'Tin God' in Morgan slang. In 2010, just after completing his training, he joined the development department and contributed to the design of the new
3 Wheeler. He even built the first four bodies himself.

Richard would have loved to stay there, but he was assigned back to the tin shop. It's fascinating to watch how deftly he wields what are normally considered crude tools. It's as if this is an inherited skill – all the more likely as he shares a hobby with his father: restoring classic cars. The elder Harris owns a respectable collection of old Jaguars, Austin-Healeys, and Morrises which demand continuous maintenance and restoration.

The press brake

The red monster looks like an intruder in the tin shop. It has only been there for two years. Before that, the shop looked like one would imagine a metal working shop from an earlier era. Although individual work stations had long employed air-powered nailers, sanders and drills, those were the only concessions to modern working methods.

Sheet metal bending, too, was done by traditional, manual methods. Those machines are still to be found at their accustomed places, and are still in regular use for smaller tasks. And not every one of the aluminium pieces, each a minimum of 1.5mm (0.060in) thick, is individually cut. Nowadays, this is done by suppliers equipped with modern machinery, such as laser cutters.

The flat, cut-out parts are stored in a large rack directly next to the press brake. For the past ten months, this has been operated by David. Before this, he had worked in the tin shop. Like the others in that department, he had been engaged in applying aluminium sheets to the wooden frames, until issues developed in both elbows, and he could no longer wield hammer and pliers without pain. Now he's happy to be working in familiar surroundings, at a new work station.

The "red monster" is a modern press brake, installed just two years ago.

Introduction of the machine has once again resulted in a significant quality improvement, because there are simply some work processes that a machine can do more accurately than even the most skilled craftsman. David has precise cutting plans for every part as well as specifications for every bend angle. These he inputs to a display screen. Today, he is responsible for all extensive bending operations in the shop. Moreover, every work station in the tin shop has a special box, packed by David, containing all the necessary parts.

David operates the machine ...

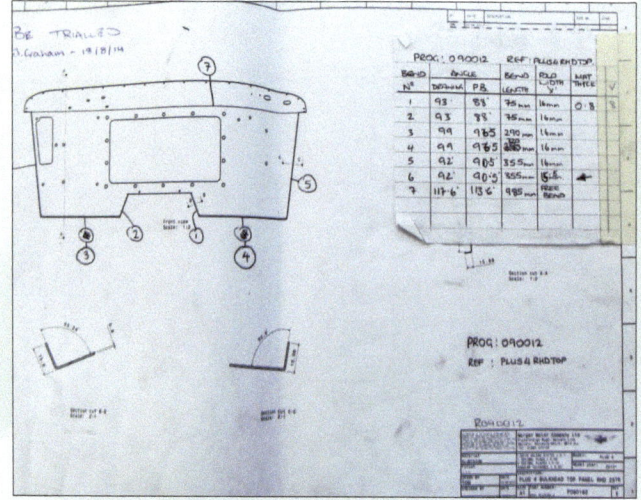

... and has a blueprint for every component.

Sheet metal shop – second day

Richard fits the scuttle panel …

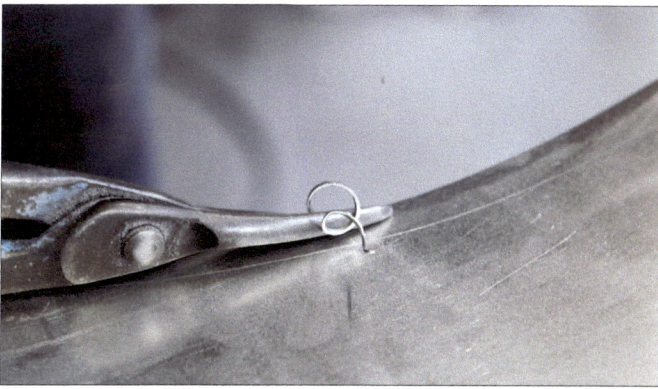

In skilled hands tin snips can do amazingly fine work.

… a great deal of finesse is required to bend its sheet metal skin into shape.

Again and again, surfaces are carefully filed.

Meanwhile, 'our' car has indeed been given priority status. To win back some of the time lost in the elbow drama, Edward assists us with the chassis. He has been with Morgan for six years, and alternates between the tin shop and the assembly shop. He starts with skinning the driver's side elbow, this while Richard is still working on the second door; its delicate surface is still protected by a wooden template while it is being fitted. After the bottleneck, the tin shop is again filling up with new wood frames, to the point where they are piling up in the corridor.

After Richard has completed both doors and fitted them with precision, he turns his attention to the elbow skin on the passenger side. Meticulously, he hammers the edges to achieve perfect shaping. Next, the panel is marked to indicate how much excess material is to be sacrificed.

Naturally, there are no less than four different snips employed for this, to carefully remove even the tiniest excess. What remains is smoothed by file and sandpaper. This is followed by primer and sealant.

Before final assembly, Richard reworks the tiniest imperfections, one more time, until the piece satisfies his critical eye. Time to do one side? Two hours and ten minutes.

Normally, in the tin shop, one man will work on a frame alone, from beginning to end. Of course with two on the job, it goes faster. While Richard is still working on the elbow, Edward installs the safety belts – after first checking them for possible damage. That is done quickly enough; after all, there are just two bolts to tighten.

The weld beads are smoothed.

After priming, a generous coating of sealant is applied to the elbow.

Now it's time for the larger panels. While Richard works on the scuttle panel – the body piece ahead of and to either side of the windscreen – Edward turns his attention to the back panel. On the sport model, the spare wheel is deleted, for aerodynamic and weight reasons. In its place, the top edge of the panel has a cut-out for a third brake light, which on the normal Plus 4 is mounted on a gooseneck above the spare wheel – not the most visually attractive solution. Most Morgan drivers remove this but keep it in storage, just in case.

The scuttle panel is that piece of sheetmetal ahead of the windscreen that meets the rear of the engine bonnet, stretches downward to the running boards, and the leading edges of the doors. It is a rather unwieldy piece of metal, which first has to be bent in two places, by hand and by eyeball alone.

After that, it can be attached, at least temporarily, so that the cut-outs for the door hinges can be made. Protected by close-fitting templates, it is first fitted, and the excess material is cut away with tin snips. Then begins the careful process of bending, crimping and filing, until the panel fits perfectly. Further work is done at the workbench.

Edward proceeds with the back panel in the same way, except that he has already bent the straight sections using a press. The back panel requires one extra step: The undersides of the end pieces need to be welded together. Each worker does his own welding in the welding booth at the rear of the tin shop.

Like all edges, the weld beads are carefully smoothed. Installing the back panel is business as usual. In the middle of this operation, the signal announcing quitting time and the start of the weekend sounds, but Edward would really like to finish this stage of the manufacturing process. He crimps, files, and sands for another half hour (of overtime), until the surface of the back panel meets his personal, demanding standards.

Edward installs the inertia reel safety belts.

The first work week – our story so far

We have reached the end of the first of three planned work weeks. The chassis was assembled in the chassis shop, with all of the parts required for operation on the road fitted. It now has an engine, suspension, wheels and brakes, as well as several elements to allow mounting of the wooden frame and bodywork. Now the body – consisting of the wooden frame, the aluminium surfaces attached to it, and the doors – will be attached to the chassis in the tin shop.

The delay caused by the elbow allowed us to see how 96 shaped wooden pieces came together to make a frame, and were skinned in aluminium to create a body. This job is nearly done, ready to be handed over to the assembly shop, where this and other bodies are eagerly awaited. It will be just a few hours more.

Try this yourself: Richard forms a wire bead edge.

The scuttle panel

There's really just one more part that needs to be worked, but as always, the devil is in the detail, and this will delay us until lunchtime. The first task is to bead a wire into the inside of the scuttle panel. This is tedious work. Richard wraps a hammer with crepe paper to avoid damaging the delicate aluminium. Next, the leading edge has to be given a z-shaped jog, so that the engine bonnet (hood) can later rest on the resulting strip of metal and be flush with the top of the scuttle panel. Martin helps Edward at the machine.

Using a template, holes for various cut-outs are marked. Later, the three wiper spindles, characteristic of Morgans, will be installed in these. These triple wipers may look toy-like, but they perform very well. Punching the holes is also done with a positively ancient machine.

The scuttle panel

Forming the edge is a whole-body experience.

The roller forms a 'jog' in the metal panel.

The padding on the hammer protects the aluminium surface.

For longer sections, Martin lends a hand.

Richard punches the holes for the three windshield wiper spindles.

The bonnet (hood) will rest on the front lip. Repeatedly, the panel is laid on the frame to check its fit.

Edward joins in.

Richard in the welding booth.

The wire reinforces the door frame.

The crash pad – the padding around the dashboard end of the scuttle – is fitted.

Repeatedly, the surface is checked for flatness. And repeatedly, corrections are made, checked, and corrected again. Richard lends a hand to Edward – at Morgan, everybody helps everyone else, quickly and as a matter of course, to avoid any idle time.

On the underside, the metal is reinforced to take the windscreen mounts by bonding in a metal doubler. To accomplish this, the metal surfaces are roughened and joined by a two-component adhesive that cures in just minutes.

Next, Richard ducks into the welding booth for a moment to weld the edges, which are sanded down with great care.

The scuttle panel

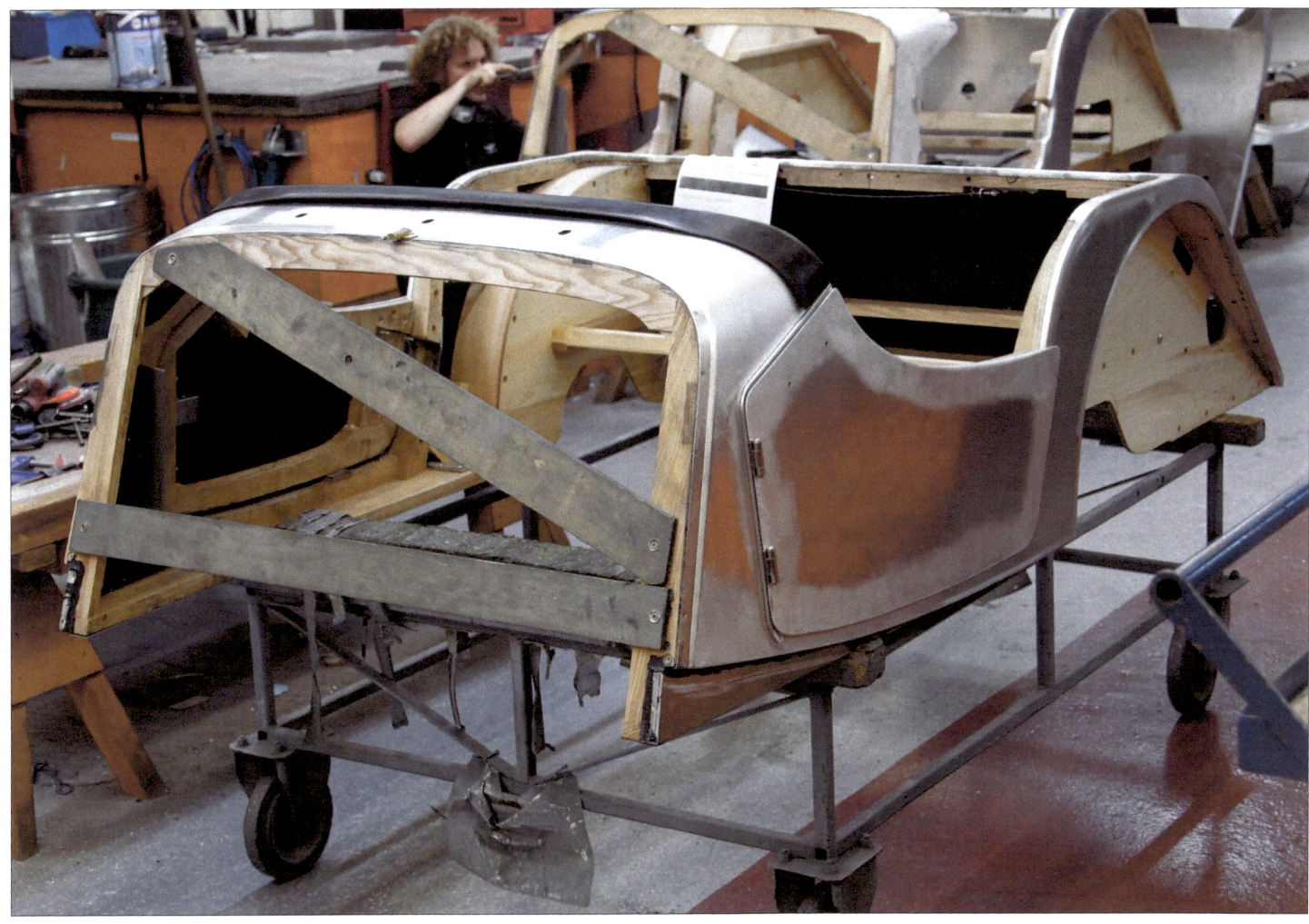
Work in the Tin Shop completed, the frame is now sheathed in metal.

In the next step, the insides are given a generous coat of primer and the joints are sealed, before the scuttle panel is attached using tiny nails. An air nailer is used, at locations that will not be visible later.

A long piece of wire is also glued to the bodywork. This wraps around the leading edge of the door like a rain gutter, to ensure that no water enters the gap. Not to worry, the adhesive bonds firmly to metal – as Richard assures us, 'like shit to a blanket.'

The last item in the assembly kit is the so-called crash pad, which seals the bottom of the windscreen and also – hence the name – pads the dashboard in the event of a collision. After cutting to shape and attaching, the only things missing are holes for the outside mirrors on the doors, and no small amount of detail work with hammer, file, and sandpaper, until everything is truly perfect and the fit is optimal.

Finished! Edward helps Richard set the body on a dolly, and push it to the ramp leading into the chassis shop. The tin shop completes its work at 1:15pm. Total time? 12 hours and 10 minutes, with a major contribution from Edward to speed the process.

Export model. This 3 Wheeler is going to Australia.

A digression: production of the 3 Wheeler

Mark Cerrone.

Even though this book describes production of the traditional models – 4/4, Plus 4, and four-seater – we want to briefly turn our attention to the 3 Wheeler. The Aeros are not much different from the traditional models – they all go through the shops in the same sequence, and are largely built by the same workers.

Not so the 3 Wheeler. This is built in its own dedicated shop. The chassis is supplied by ABT of Ross-on-Wye, and the body by Premier Sheet Metal (Coventry) Ltd. The frames are painted in the paint shop, after which the car is assembled by one of the dedicated workshop's seven workers.

Each worker assembles a single car, from beginning to end, which of course results in enormous pride in the finished product. In conversation with Mark Cerrone, head of the 3 Wheeler department, I suggested that each vehicle carry a plaque with the name of the worker who assembled it. He said he would consider it.

On average, it takes a single worker one week to produce a 3 Wheeler – including small special-order items such as extended, padded headrests for the Australian market.

A 3 Wheeler in the paint shop.

The snug interior of a 3 Wheeler

A finished 3 Wheeler on factory grounds.

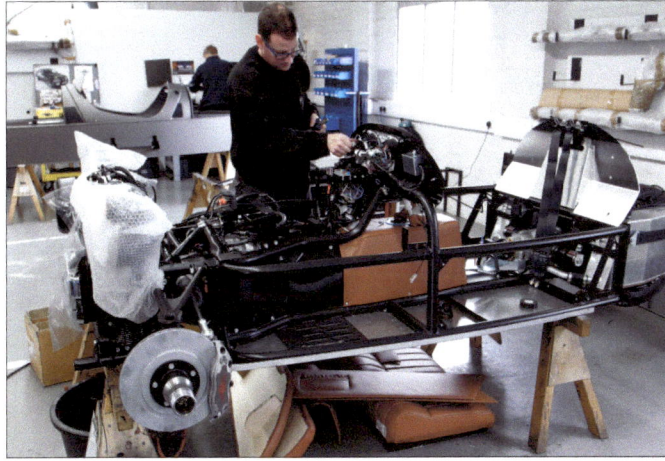
One man, one 3 Wheeler. Every 3 Wheeler is built in its entirety by a single craftsman.

The 3 Wheeler hall.

The 'marriage.' The body is joined to the chassis.

Assembly Shop

The assembly shop is teeming with cars, lined up and awaiting their turn. This is still the aftermath of the elbow drama. Luckily, our car's 'VIP status' once again helps it along, and so just after two o'clock, Jason takes 'our' Morgan under his wing, as Karl will go on a vacation the following day.

In the assembly shop, the body will be mounted on the chassis. Then fitted with wings (fenders) and bonnet (hood); also the electrical items are pre-installed. Because Karl has already installed the fuel tank and seals between chassis and body, we are on the brink of the most eagerly awaited moment in the manufacture of any car – what you might call the 'marriage'.

Except that term is not used at Morgan, and the fit proceeds in a decidedly unspectacular manner: two men, four corners – Karl and Jason simply lift the body from its dolly and precisely locate it on the chassis. That's it. No bells, no minister, the eternal bond (at least, for the life of the car) is completed in the blink of an eye. The body fit is adjusted by a few wedges and wooden shims, and then joined to the chassis by fourteen bolts. Time required? Not even fifteen minutes!

In Morgan parlance, this is a 'roll bar.' It adds significant reinforcement to the knee area.

Jason Hallett – the golfer

(sheet metal at work, irons on the links)

Jason is 45 years old and – like so many Morgan workers – a native of the region. At age 16, he started his five-year apprenticeship as a sheet metal worker at the factory. That was 29 years ago. He has been at the same work station that entire time. We don't know if his father encouraged him to apply at Morgan, but in any case his father had his own long-term history with Morgan, in the mill. Of course that was some years ago. Jason is divorced and father of a 17-year-old daughter. In his spare time, he likes to play golf – his handicap is 24.

Fourteen bolts hold the body to the frame.

MAKING A MORGAN · DAY 6

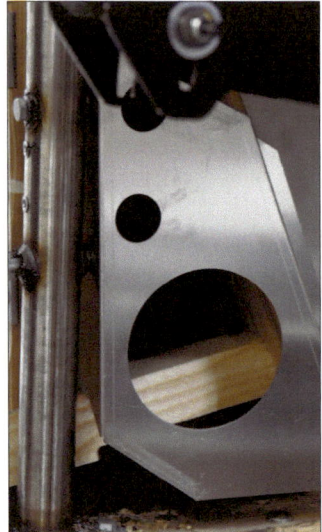

The audio speakers are discreetly installed in these cut outs deep within in the footwells.

Let's see here ... how wide is this wing supposed to be?

At the front, screws attach the wooden frame to the bulkhead.

Next, the so-called heel board is bolted in. This is a wooden board to separate the interior from the differential and battery box. Then it's time for the roll bar. This is not, as one might assume, a protective measure for the event of a rollover; it is instead a reinforcing hoop installed in the footwells.

Nevertheless it makes a significant contribution to passive safety, as it protects the occupants' legs in the event of an accident. The part is made by an outside supplier, and, along with other parts (body mount kit, red wiring kit, and blue engine kit) is provided to the individual work stations on a pre-packed trolley by the goods in – goods out department.

Next, the bulkhead is joined to the body. Karl has already mounted the seals, so it only remains for the bolts to be installed. Tightening the bolts forces out excess sealing material, which is trimmed away. This whole process takes a matter of just a few minutes.

By contrast, installation of the next component is a bit trickier. The fuel lines need to be securely fastened to the chassis. Not a big deal in itself, but the lines are installed to the chassis from below, which calls for a bit of fussing around.

Kevin has a much easier job of it, installing the speaker buckets. These housings for the sound system speakers are mounted to the roll bar, far forward in the footwells, so that their appearance does not detract from the puritan look of this automotive classic. On our own personal Morgan, we renounced newfangled gadgetry, such as a radio. Even the sight of an aerial would be offensive to us, especially since the sporty sound of the engine was music to our ears. Still, many other customers don't deny themselves such minor modern luxuries.

Now, we get down to brass tacks. Jason turns his attention to the rear wings (fenders), supplied by Superform Aluminium of Worcester. A single part is used for all traditional models, custom fitted in assembly. In our case, Jason bolts the rear wheels in place, and measures the exact gap between the inner edge of the body and the outside of the tyres. This takes into account the various possible tyre sizes.

'Superform' is no exaggeration. The presses are enormous.

Dave Edwards.

A visit to Superform Aluminium in Worcester

Superform Aluminium and Morgan have been working together since the late 1990s. Our host at Superform, technical director Dave Edwards, is partly responsible for the relationship. As a young designer, he had worked for Morgan as a freelancer, developing, among other things, the new hood for the traditional models. In the event of rain, this can be raised in a matter of seconds, from inside the vehicle, or unlatched and flipped up to allow comfortable loading of the storage area behind the seats.

Even while working for Morgan, he had also undertaken projects for Superform Aluminium. Founded in 1974, the company is mainly a supplier to the aerospace industry, and specialises in hot forming of aluminium and magnesium.

Charles Morgan had long been searching for ways to (cautiously) streamline the manufacturing process in his small car factory. Because forming wings by hand was especially time consuming, and carried with it the danger of dimensional inaccuracy, a great deal of thought was given to a co-operative approach. That is how Superform got its first major automotive contract. Today, besides Morgan, the prosperous Worcester firm is the go-to supplier for most of the limited-production luxury automakers, including Rolls-Royce, Bentley, Aston Martin and Ferrari.

MAKING A MORGAN · DAY 6

The sides of the ten-tonne press tools hint at the final shape of the wings.

This is how the finished wings (fenders) arrive at the Morgan factory.

A lorry brings a new load of preformed wings to the factory.

Their co-operation began with the manufacture of wings for the 4/4. Naturally, Superform could have delivered perfect wings, with no dimensional deviation, but in the case of Morgan, this was not in keeping with the company philosophy. The parts should be identical, but with a certain tolerance between left and right wings – small, calculated irregularities so typical of genuine hand craftsmanship. This is characteristic of Morgan, and had to be maintained, even in manufacturing on an industrial scale.

Meanwhile, Superform also makes all Morgan wings, not just for the traditional models, as well as the cowl at the front of the car, and inner reinforcements – valances. The front wings (fenders) and both valances are made on the same tooling and in a single manufacturing process. A number of separate forms are used for the remaining parts. The tool for the simultaneous forming of the four parts weighs more than ten tons alone, and has been in service, without interruption, since the beginning of the intensive co-operative effort with Morgan.

The production method is unique as well as fascinating. A massive sheet of aluminium, 2 millimetres (0.80in) thick, with an area of about seven square metres (75 square feet) is coated with graphite and then placed in a gigantic 1600 ton press. In a complex, 30-minute process,

A visit to Superform Aluminium

Incredible as it seems, trimming the wings is done freehand – thanks to a wealth of experience.

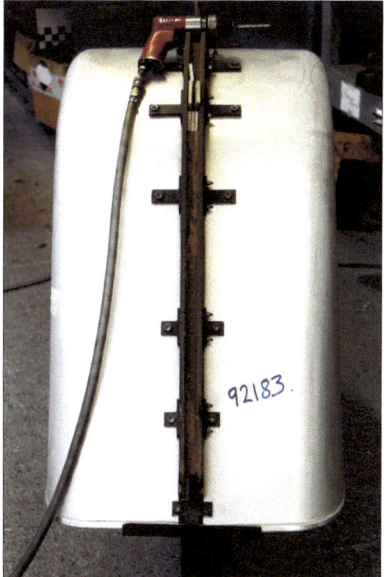

This drill template ensures that wing mounting hole placement is perfectly symmetrical.

The order number is stamped into the wings.

the sheet is heated to 480°C (900°F) and shaped into the desired, three-dimensional form. It is also possible to form significantly thicker material. We saw 6mm (1/4in) plates in storage, and heard of far more impressive capabilities. The currently possible maximum metal thickness is 'top secret.'

In the final step, the edges of the graphite-blackened parts are accurately cut out by a laser cutter, and the pieces are dipped in a phosphoric acid bath to regain their metallic lustre. Incidentally, the front wings for the 4/4, Plus 4, and Plus 8 are identical. They are trimmed accordingly at Morgan, and individually fitted to the cars. There is still a call for hand craftsmanship.

Using a metal-cutting saw and pliers, a piece is cut out of the end of the wing. Then the desired width of the wing is marked, and the excess material is easily cut away with powerful tin snips. I held my breath, but Jason went about this operation confidently, and without a hitch.

Now, a little more filing, and the job number is stamped inside the wing. Then both wings are clamped in a jig that serves as a template to drill the necessary holes. But first – it's quitting time again!

Day 7

The wings are bolted in place.

A massive metal jig ensures perfect fitment of the front end sheet metal.

Looking like a car now

Until now, I had thought that work times at Morgan were sacred. I was quite mistaken; when I arrived at the assembly shop at the usual eight o'clock, I found that Jason had already been at work for an hour, and had fastened the rear wings by means of their eight bolts.

Now it's time for the front Superform wings. Before he fits them, the still somewhat unstable front of the future Morgan is reinforced by a massive assembly jig, which will be removed later.

In assembly, the narrow and wide bodies differ only in the amount of metal cut away from the 'universal' wings supplied by Superform. The wing fit is tried, again and again. Jason cuts the large, straight sections with electric metal shears. When millimetric accuracy is called for, he pulls out the manual tin snips.

I was allowed to try this for myself, parting a piece of cut-off scrap with the tin snips. The force required is enormous, as the wings are appreciably thicker than any

Looking like a car now

Try it yourself – fine metal tailoring of two-millimetre aluminium sheet.

Before the front wings are bolted into place, holes need to be drilled.

Before and after. The wing on the left has been trimmed to size.

of the materials used thus far. But Jason makes it look like he's cutting cloth.

Before attaching the wings, thin strips of leather are glued onto the valance to eliminate the possibility of rattles. Then Jason welds plates with the job number to the ends of the 'wings.' The wings are attached with ten bolts each. Once additional reinforcement, in the form of top and bottom wing stays, is added, the wings achieve the required stiffness. Now the jig can be removed.

Next, another component supplied by Superform – the nose cowl – is fitted with a rubber seal and bolted in place. The cowl gives the front of the car its characteristic shape and will later house the radiator grille. Strips of flexible woven material – 'bonnet lacing' — are installed to prevent the engine compartment bonnet from rattling against the cowl. Gradually, the grey parts box is emptying. Now it contains the windscreen washer fluid reservoir, whose mounting bracket is riveted into place. Next comes the propshaft casing, the rear cover of the propshaft.

The car gets its face; finally, the nose cowl is mounted.

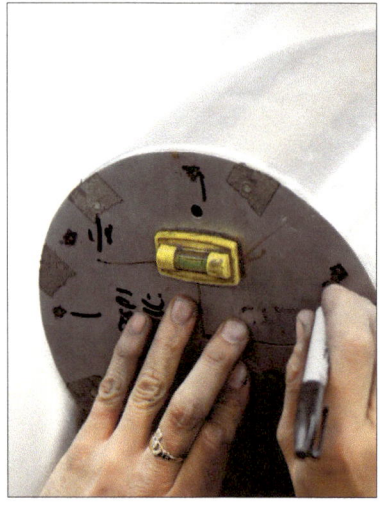

Thanks to a template with a built-in spirit level, the headlamps will ultimately be mounted perfectly level in their buckets.

Anything that isn't perfect, goes back to the previous station for rework.

In drilling the holes for the tail lamps, Jason has indeed managed to place one hole in the wrong location. "Shit happens," he says. He immediately marks the spot and puts a note on the wing that he'd like to have it back, briefly, as it has to be unbolted by the paint shop (the next department) anyway. The repair on the following day takes only a couple of minutes.

For now, the drilling continues. There are many holes, large and small, for things like the hood latches, turn signals, and headlamps – these are located using a special jig with a built-in spirit level.

At 1:10pm, Jason has completed his work. The car is lifted from its stands and turned over to colleagues in the same department.

Not ten minutes later, it's back to work: Blake and Marcin, so-called 'bonneteers,' fall upon the car and begin to fit its bonnet. This is pure hand craftsmanship. Starting with a flat piece of aluminium, the wonderfully curved bonnet, with its characteristic louvres, is created.

As a first step, Blake and Marcin take both bonnet parts to an old machine that bends, edges and folds, in a single operation. Blake does more work on the bonnet. A rail is fitted to the fold at the top edge; it will soon fit into the bonnet strip, allowing both halves of the bonnet to be opened. This is a very protracted, complex process, in which repeated corrections are made by means of a hammer.

Meanwhile, Phil has taken the remainder of the car and ensures that electrical current will flow through the

Looking like a car now

The bonnet (hood) gets its elegant curves through hand craftsmanship.

Marcin Olszewski – the Bonneteer

Marcin is 31 years old and was born in … no, in his case, it's not Malvern, but rather Grudziadz, Poland. As such, he is a genuinely 'exotic' member of the Morgan workforce, as all the others are all born-and-bred Britons. While Marcin was studying quality management in Poland, he wanted to try working in England for six months to help pay for his studies.

In England, he heard of Morgan and was immediately fascinated. But in his first interview at the factory, he was turned down. After nine months back in Poland, he made a second try at Morgan and was hired. That was in 2011. First, he spent two years in the tin shop, before transferring, at his own request, to the assembly shop. He has been a bonneteer for 17 months. Marcin's hobby is astronomy, he likes to listen to music, and is interested in anything and everything to do with cars. He is married and is the father of a four-year-old, with another on the way.

Blake works on achieving a neatly installed bonnet hinge strip.

completed vehicle – at least, he lays the foundation, by installing wiring harnesses, provided by an outside supplier, and routing the wires to the correct locations. The first harness leads to the rear, and supplies electrical current to the fuel level sender and the tail lamps, among other things.

Blake has returned to the car with the pre-bent bonnet half, ready for installation. First, he sizes up the situation. As in every metalworking operation at Morgan, the tin snips soon come into play. Once satisfied with the results, he returns to his work station and carries on – drawing on his depth of experience and a practised eye.

The next wiring harness that Phil pulls out of the box is of monstrous proportions and first has to be untangled. The wiring sets of all Plus 4s are identical, regardless of customer

Looking like a car now

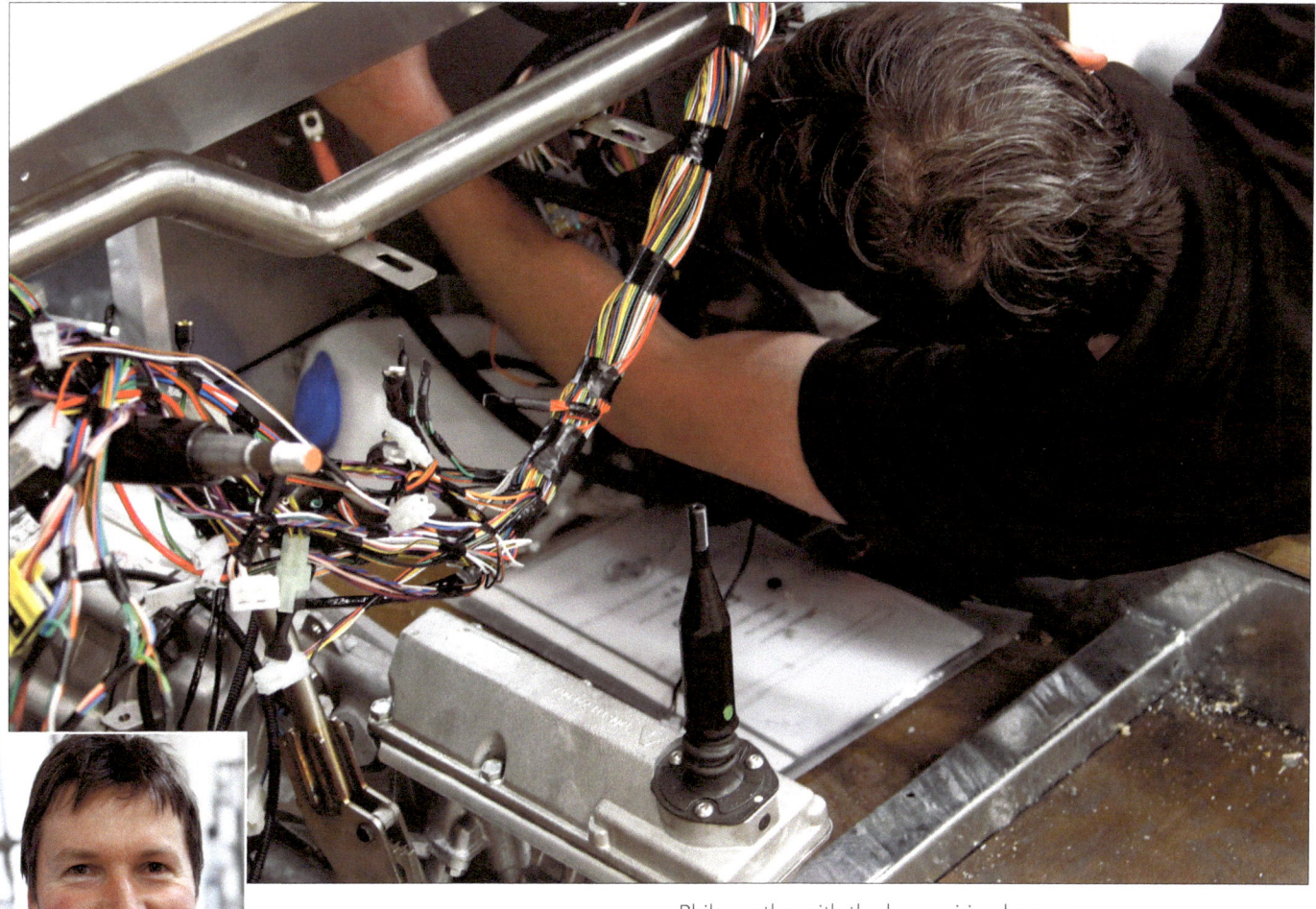

Phil wrestles with the huge wiring harness.

Philip Jones – the Electrician

Phil is 46 years old and hails from the next major town, Worcester. After finishing school, he trained as an automotive mechanic and worked in several auto shops. He joined Morgan in 1989. He is most fascinated by Morgan's uncompromising craftsmanship and authentic British roots. At first, he worked in the warehouse, but after 13 years he wanted to work with his hands, and asked for a transfer to the assembly shop. Today, he does all the tasks in the shop, with the exception of the engine compartment bonnets (hoods). Phil is heavily involved in sports – athletics and physical fitness, bicycling and hiking – and is an enthusiastic motorcyclist.

options. They are carefully checked and, if necessary, the insulation of individual wires is repaired. It looks like mass confusion, but gradually, order comes out of chaos as the first branches are laid into their appointed places.

Now the battery is connected to the wiring harness, and one of the first load items is installed – the horn. Actually, Phil wanted to install the 'wiring jar' – a sort of main junction box where most of the wires terminate, and where the fuses are located. But Blake is still occupied with fitting the bonnet, so Phil continues working on the interior.

And before you know it, it's quitting time again. I was about to pack up my things when Mick told me he wanted to continue working on the car. After the shop had emptied, he leisurely fitted the windscreen and sidescreens.

The fittings for the sidescreens are attached to the doors.

Mick test fits the sidescreens.

Using four bolts, Mick attaches a temporary windscreen mock-up and checks the angle of inclination. Ideally, this should be 63.0 degrees. And … it's 63.0 degrees.

Next, he attacks the sidescreens, whose brackets have to be installed first. It's important that both side windows have the exact same height. Overall quality of fit has supposedly been greatly improved since 2013.

In the case of our own 2011 Plus 4, the front edges of the sidescreens aren't an exact match to the windscreen frame, which causes occasional problems. But in this case, the angle is perfect, a fact that I'm able to confirm for myself.

After multiple checks and re-checks, Mick drills the holes for mounting the windscreen. After a little less than an hour of overtime, Mick calls it a day. After all, the boys in the assembly shop will all be back, bright and early, at 7am the next day.

Day 8

Connection works

For us, that means no leisurely breakfast in our hotel, because that isn't served until 7pm. Instead, we march off to the factory. All are at their posts, right on time, and work continues unabated. First, Mick has to make a new crash pad, as the one made by Richard on the previous day was urgently needed for another car.

Marcin takes over from Blake on the engine bonnet, and rolls a reinforcing steel wire into the rear. Now there are three people working on the car in parallel, and work proceeds at a rapid pace.

At 7:50am, we achieve a highlight: The car is pushed toward the shop door, and parked in 'pole position,' as it were – effectively with one wheel in the paint shop. Now Paul installs the throttle (gas) pedal – even at Morgan, this too is a 'fly by wire' component, as on so many other marques. The brake light switch is installed ahead of the brake pedal.

After Marcin has re-adjusted and refitted the bonnet one more time, the electrical distribution box is installed. This is sealed multiple times to ensure that it is water-tight. Most wires end in this box. The electricals now start to look almost neatly arranged, if that concept can be applied to the multitude of individual wires and small components.

Mick takes charge of the blue engine kit parts box, containing the components for the next assembly steps. For the first time in the life of this car, our Morgan

Mick fabricates a new crash pad end.

The blue box contains the engine compartment kit.

The electrical distribution box is put to its intended use.

is vacuumed to remove all chips and traces of the manufacturing process. Now it looks decidedly well-tended.

Next, Mick attaches the handbrake cable alongside the propshaft tunnel, before routing a thick hose in the engine compartment. This so-called intake assembly directs outside air through a filter to the cabin heating system.

Mick blows out the fuel line and attaches it to the engine. In giant steps, we are reaching the final stage – a drivable car. Since we're dealing with lines and hoses, the windscreen washer hoses are attached, followed by the coolant reservoir and its connection to the engine cooling system.

Now Mick installs the gearbox cover. The transmission gear (shift) lever emerges directly from the gearbox cover. Mick has to crawl deep into the car to attach the bolts behind the throttle pedal. On a modern automobile assembly line, the entire auto body would probably be upturned to allow the worker to install the part, in the optimum way as determined by time-and-motion experts. On a modern line, this would take seconds. But we're happy that this isn't the case at Morgan. We love the marque, for this build approach and so many other reasons.

Mick takes a heater from the stack and mounts it in the engine compartment, against the bulkhead. With great

Connection works

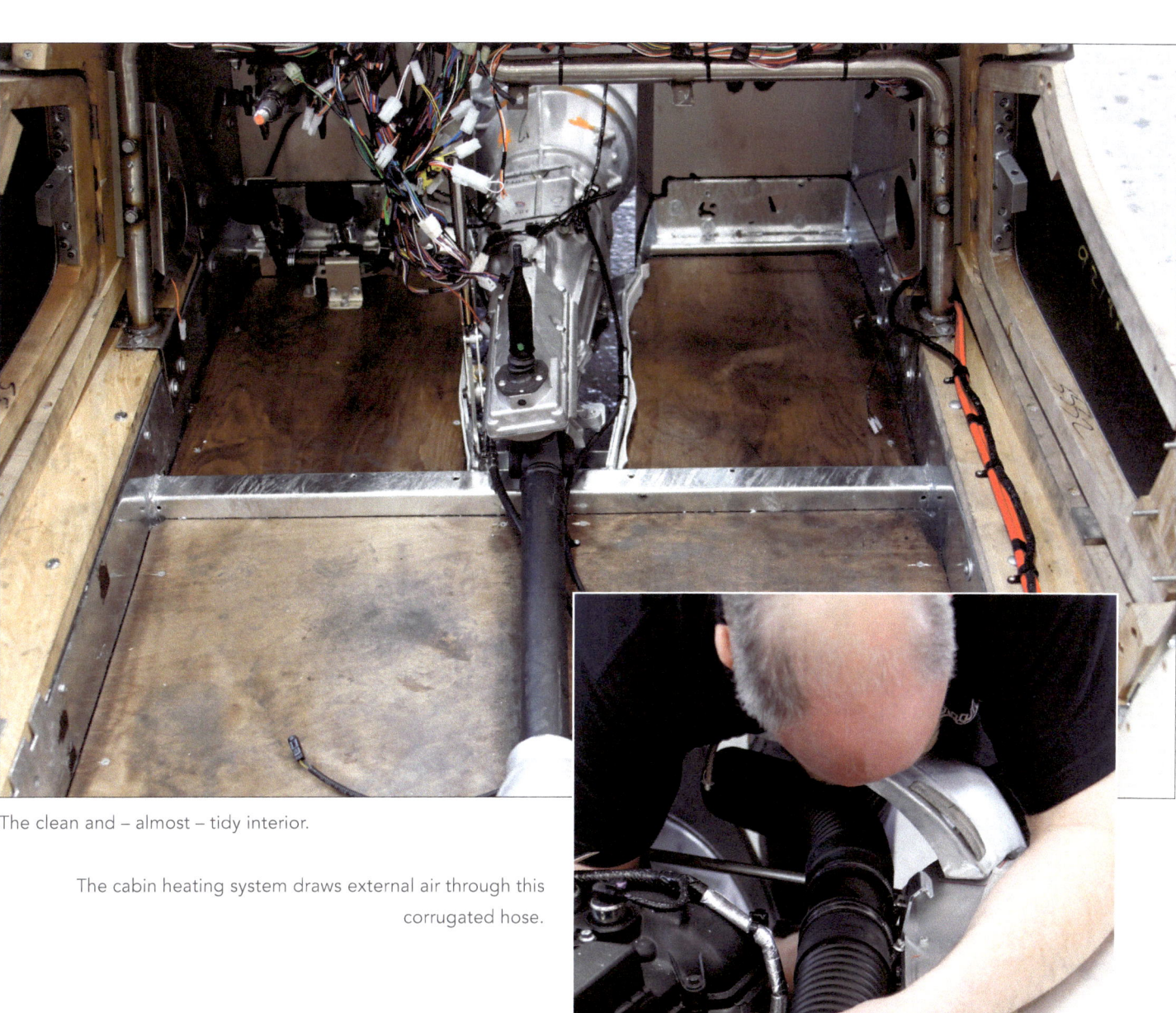

The clean and – almost – tidy interior.

The cabin heating system draws external air through this corrugated hose.

concentration, he seals both sides of the bulkhead – several times, just to be sure. This, among other things including the fit of the sidescreens, shows how Morgan's assembly quality has gotten perceptibly better in the past few years.

A change of scene. Let's go back to Marcin. With an assistant, and the skill born of experience, he now punches the characteristic louvres in the bonnet, forming a perfect arc ... 27, 28, 29 ... finished. Without careful counting, it would be easy to make one too many – or too few.

Now the protective film is peeled away from the aluminium sheet, and the bonnet looks like that familiar,

The bulkhead is sealed. The heater box has already been installed on its forward side.

With a good deal of contortion, the gearbox cover is installed.

beloved part so characteristic of a Morgan. A bit of final adjustment – some bending here, a tap of the hammer there – and it fits. Then it's removed again, so that Mick can continue his work in the engine compartment.

Part of the propshaft is still visible, and needs to be covered. Now that I'm more familiar with the English designations, I guess that this is called the 'propshaft cover' – and I'm right. Before installing, Mick applies self-adhesive rubber strips to the ends of the adjoining elements to eliminate any possibility of rattles. The upper of the two wires goes to the

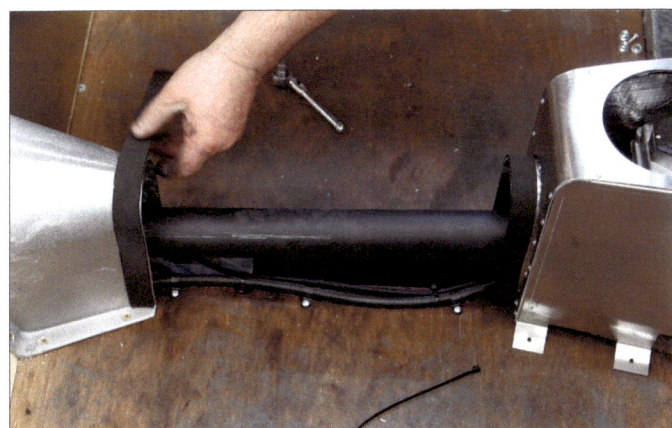
Before the propshaft cover is installed, Mick sticks rubber strips in place to prevent rattles.

Just don't lose count – 29 louvres are punched into each bonnet half.

Mick Bishop – the athlete

Mick is 58 years old and a native of Worcestershire. He had spent his entire career in automotive repair, most recently commercial trucks – before joining Morgan ten years ago. In his late 50s, he still engages in sports. Until recently, he was still an active soccer player, until an injury forced him to lay to rest that part of his career. Now, he is an impassioned motorcyclist.

Mick is married with two children. We have already met one, without realising it. The next day, he asked me if I had recognised his son on the shop floor, but I had to admit that I hadn't. It was Blake, who happened to be doing his apprenticeship in the same department where his father works.

safety belt buzzer system, which checks whether the seat is occupied and the belt buckled.

After that, Mick assembles the air cleaner and installs it in its appointed place. Then, the activated charcoal filter of the fuel tank venting system. This component is mounted above the tank and again requires Mick to contort himself. But at least this keeps him supple.

One of the hoses attached to the tank is stretched extremely tight, but Mick can't take care of that himself. He notes the problem in the build book and so passes it to the final finish department.

The last item on Mick's list is connecting the air filter to the engine and a final vacuuming. Now the car is ready for

It is … a car! Kevin rolls it out of the assembly hall.

Kevin Bunn – the career man

Kevin is 36 years old and hails from Worcester. At 16, he started in the tin shop, after his father had sent his application to Morgan. At age 25, he was not only appointed foreman of the tin shop, but also took responsibility for the sheet metal and welding shops, with a total of 26 workers. His brother also works at Morgan. Kevin is engaged and does not have any children yet – but two dogs keep him very busy. In his leisure time, he follows typically English pursuits – gardening and football.

final inspection by Kevin. The clock on the shop wall says 12:15pm. It also tells us that through great effort and quick work, the crew has made up for half a day of lost time.

Final inspection takes place at one o'clock, and Kevin can't find anything to fault. Just 15 minutes later, he pushes the car over to the paint shop, whose entrance is blocked by an already painted Aero coupé.

Paint Shop

The Morgan doesn't stand there long. The Aero is soon rolled away, and a half hour later, Chris, in the paint shop, takes over. He pushes the car into the shop and begins by removing the bonnet. Then the car is jacked up and partially dismantled; wheels and wings are removed. A dolly is bolted to the axles; any overspray on the dolly is of no consequence. Now Chris begins to coat all edges with waterproof adhesive, a light-coloured paste that will protect the body joints from moisture. Not much more is done on this day.

Paint Shop

92183 has to wait its turn; there's no free space in the Paint Shop.

In the Paint Shop, the expensive wheels are temporarily replaced by a wheeled dolly. Unlike the real wheels, any paint overspray on this structure is of no consequence.

MAKING A MORGAN · DAY 8

The wings have hardly been installed before they are again removed. Chris detaches the wings to prepare them for painting.

The colours of Morgan – an agony of choices

When we tell people that we drive a Morgan, we are often asked if it's green. Apparently, many people expect a British roadster to be painted green – more precisely, British Racing Green. As if there aren't any other colours in this world …

In the era when that most sacred of all British marques, Rolls-Royce, reported the horsepower output of its cars as 'adequate,' its upper-class clientele took for granted the fact that one did not drive the Rolls – one was instead driven by the hired help. The customers had no need to concern themselves with such mundane matters as horsepower output. 'Adequate' would also be the understated, standard reply when the question of Morgan colour choices arises – it is a breathtakingly huge palette.

We will attempt to illustrate this with a hypothetical example. Imagine a line of traditional Morgans, differing from one another only in their shade of green, lined up, bumper to bumper, beginning in Times Square, New York. The line would easily reach 16 miles in length. Now think of all the other available colours – well over 40,000 colour choices – and the line of imaginary cars would extend for more than 90 miles. If Morgan were to actually build all these cars, its production capacity would be booked out for the next 40 years.

Naturally, there are a few standard colours, including of course a particular shade of green. But why shouldn't an enthusiast treat him- or herself to a special colour – perhaps one that was seen on a rare car? They would simply need to find out the name of the colour, and order their Morgan, in 'Lush Green L 4069' for example.

Day 9

All surfaces are once again thoroughly cleaned prior to paint application.

Paint prep work

Our Morgan doesn't stand around for long. After just a few minutes, the car is rolled to the back of the paint shop on its dollies, where it is prepared for painting by a friendly crew in white coveralls and protective masks. The room is lit as bright as day, and Adam goes to work on the wings, mounted on separate dollies. First, the parts are sprayed with a cleaning solution and thoroughly wiped dry, to remove any traces of oil from the production process.

All surfaces are once again carefully sanded with small powered hand sanders, as the paint shop has very special standards for surface quality. All hard to access areas – for example, around the headlamps – are worked by hand.

To me, after 20 minutes of intensive treatment, the first of the front wings has the feel of soft skin. But more exacting inspection under an LED hand lamp shows a few remaining irregularities, which are indicated by a special marker.

Filler is applied to those spots and, after curing, is sanded again, until all surfaces are truly immaculate. Wayne has the first wing ready just in time for the first tea time.

Our Morgan still enjoys priority status, so Ross helps with the second front wing. Soon, Adam goes to work preparing the rear wings, using the same technique. There, the irregular spots are more apparent, so that he has to

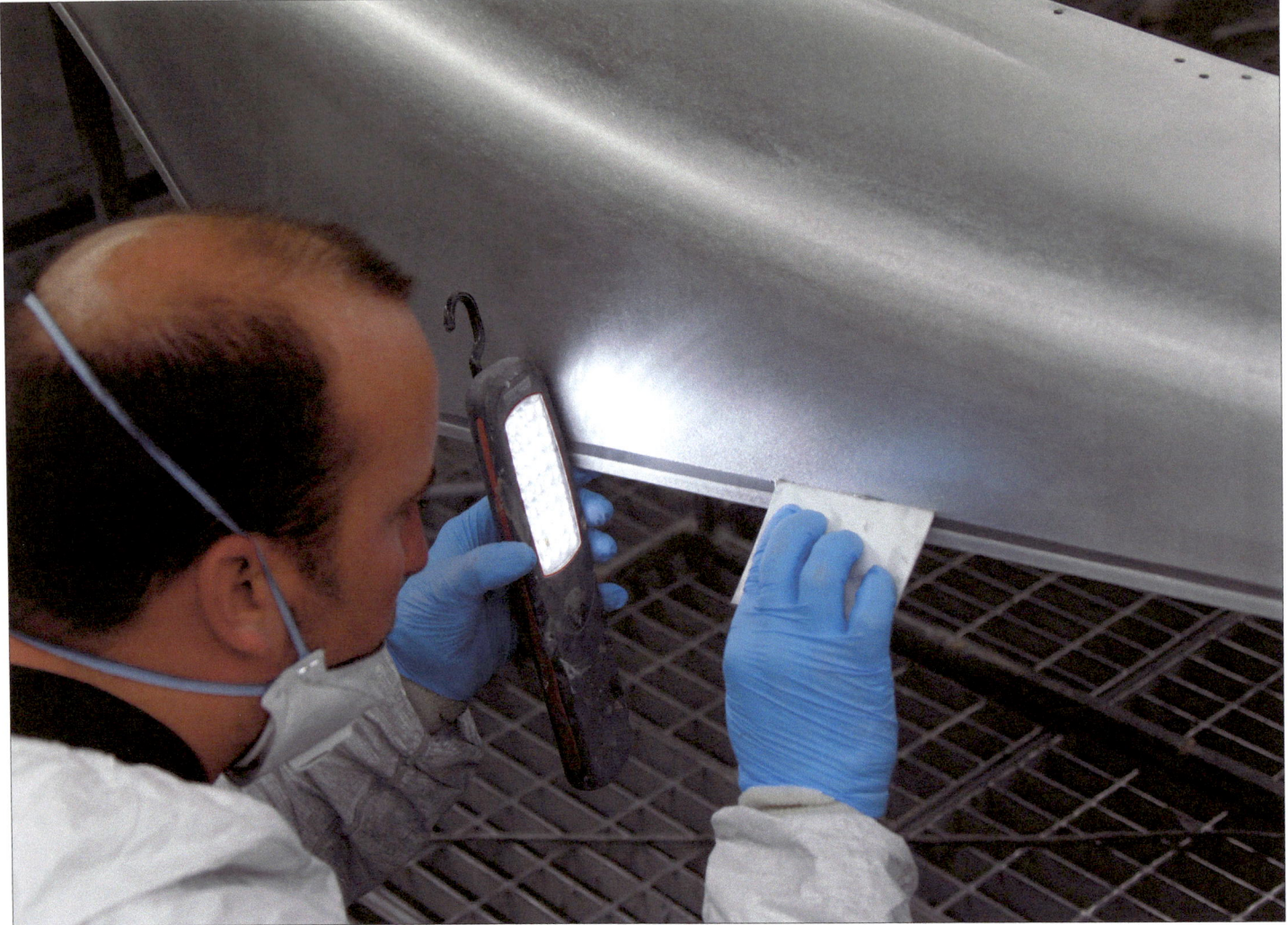

A powerful LED lamp helps detect the tiniest surface irregularities.

smooth them by carefully tapping with a hammer from the reverse side before sanding.

At 11 o'clock, the car is pushed into the neighbouring booth, where the remaining body parts are given the same treatment. First, Wayne removes the doors and lays them on a table, then the radiator cowl is mounted to a second fixture. Adam is relatively new to the firm; he has been working as an apprentice in the paint shop since May 1, 2015.

He and Wayne apply the same procedure that was applied to the wings, to all remaining body parts still attached to the car, as well as doors and radiator cowl. Only, it will take them longer. Wayne keeps a strict but loving eye on the young apprentice's work – he is, after all, Adam's father.

Not much more happens to our Morgan today. The work continues tomorrow morning, at 7am sharp.

Paint prep work

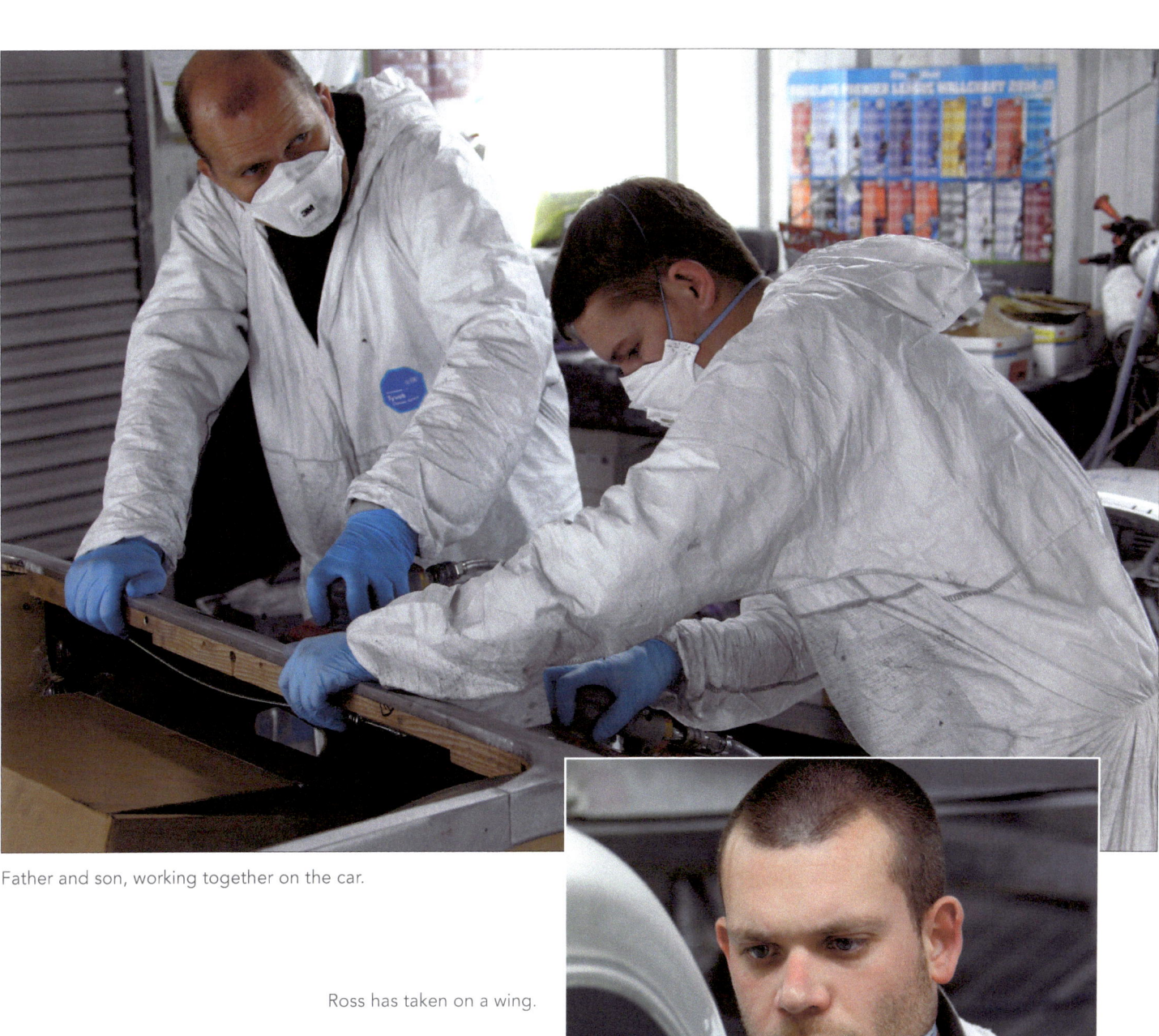

Father and son, working together on the car.

Ross has taken on a wing.

The Cole family – the 'real Morgans'

If there is one family that has been true to the Morgan works, then it is the Coles. From 1945 to 2007, grandfather Brian worked as a fitter in the assembly shop. He is now 82 and is one of the few people in the factory who has worked with H.F.S. (Harry) as well as Peter and Charles Morgan. And in his 59 years on the job, he never once showed up late for work.

Over the years, more than 30,000 Morgans went through his able hands – the old three-wheelers as well as the new generation of four-wheeled models. His second son, Wayne, born in 1965, has been in the paint shop since 1982, while firstborn Martin has been in the sheet metal shop since 1976. In the middle of the previous decade, Stacey, the youngest of the three brothers, was, like his father Brian, a workman in the assembly shop – but as an electrician.

So all of Brian's sons are, or were, at Morgan. On May 1 of this year, the personable grandson – Adam, Wayne's son – began his apprenticeship in the paint shop, representing the third generation of Coles to work at Morgan. But there are more. Since 1962, their uncle Sam has been sorting and packing parts in the goods in / goods out warehouse, and in the 1970s, uncle Peter made chips in the wood shop.

That accounts for all of the Coles at Morgan, and so the portrait of the family members standing behind the new 3 Wheeler has special meaning – and not just for the Cole family, who are visibly proud of the marque and the factory and their role in both.

Day 10

Paint Shop

Two are better than one: This is the second primer coat.

Paint Shop

The car stands ready, in front of one of the large spray booths. First, Chris wraps it up, like a package about to be mailed – at least, those parts that are not to be painted. That is most of the parts, at least where the body is concerned. Thanks to a simple but clever design, the flexible crepe masking tape and the masking paper are dispensed simultaneously.

Masking takes up a whole hour, but applying the first coat of primer takes just 15 minutes. The first layer serves mainly as corrosion protection. Now we have to be patient for ten minutes, until the primer has dried. Its shade has been chosen in keeping with the chosen vehicle colour. In a few spots, the filler used to smooth the body surfaces still peeks through, but this does not present

Chris masks everything that isn't to be painted.

As simple as it is clever: the masking paper dispenser automatically applies masking tape to the edge.

a problem. The second primer coat also contains filler, which again smoothes the surface. Thanks to a hardener, this coat is quite durable, and thinner makes for easy spray application. The colour is now a dark grey – again appropriate for the finish colour. At 10 o'clock, the drying process begins, at a temperature of 80°C (176°F). Not much more will happen today – so, off we go into the weekend.

The second work week – our story so far

After a final half-day of work in the tin shop, our narrow body was delivered to the assembly shop, where it was joined to the rolling chassis. There, the car was fitted not only with the missing body parts such as wings and the engine bonnet, created with painstaking hand craftsmanship, but also the wiring harnesses, windows, and other components such as the transmission and propshaft covers, throttle pedal, and filters.

Aside from the fact that the car does not yet have a starter, the wiring is not connected, and the throttle pedal doesn't work yet, the car is practically drivable. In the Paint Shop, all of the parts to be painted have been primed – overall, a good result for a work week in which concentrated effort and overtime were applied to meet the targeted production timeframe of three weeks. Still, there is much work to be done before the car is ready for delivery.

Down to basics: The first layer of primer is applied.

Day 11

The result of two weeks' work. Even without a high-gloss finish and final assembly, the car already looks quite impressive.

Third day in the Paint Shop

This is the third day in the paint shop. They say good things take time – and this is especially true of a good paint job. Now the task is to carefully smooth the primer. Jim goes to work. Again, much of this is strictly hand work.

The less accessible spots are sanded by hand in any case – first with 320 grit, then using a sander with 500 grit, followed by a soft sanding pad and finally 1000 grit, until the surfaces are truly smooth. Soon Ross joins the crew to lend a hand to Jim, because the paint preparation work is to be done by lunchtime. To make room in the spray booth for the huge wings, they push the completed body outside.

Both men work together on a single wing. At the end of this step, the parts are blown clean using compressed air, and then washed with a cleaning solution to remove any remaining traces of oil or grease. There follows the critical inspection using the LED lamp, and final minor touch-up work.

Now it's time for the scuttle, which is given the exact same treatment. Just before noon, all of the paint prep work is completed, and Mark has pushed the body back into the spray booth. The 'new paint shop' is now 15 years old, and encompasses three spray booths: two for entire cars along

Third day in the Paint Shop

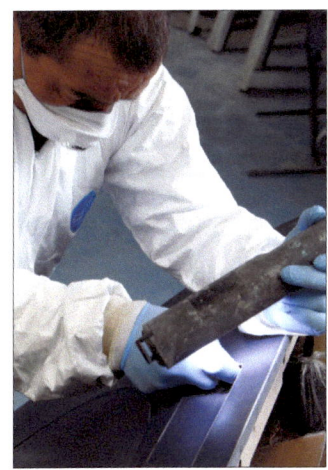

As always, the corners can only be addressed with painstaking hand craftsmanship.

Before the next painting step, everything is masked again – this time, in plastic.

with their attached parts, and one for smaller jobs such as dashboards.

As I join them, Mark is in the process of carefully packing the body. For this, he doesn't use masking paper but rather just crepe tape and thin 3M film. He works with great care. Then he gets all the remaining parts from the paint shop and sets them up in the booth.

After the lunch break, all of the parts are once again cleaned and blown off with compressed air. Mark painstakingly addresses every single drilled hole. It is truly amazing how much dust can accumulate in such tiny holes. Then all parts are wiped down with a soft cloth, blown off again, and wiped again, until they are not just clean, but immaculate.

In the spray booth, Mark once again mixes primer. This is the third primer coat, which will protect the entire body against corrosion. This is necessary because the sanding process resulted in a few bare spots. This primer is applied thin, but evenly.

Now comes the magic part – mixing the paint. But 'Lush Green,' with code number L 4069, is not listed as one of the 40,000 colours in the computer. Foreman Martin is called in for quick consultation, and has an idea. To his recollection, it must be one of the old Rolls-Royce colours, known internally as 'Velvet Green,' and used from 1961 to 1971.

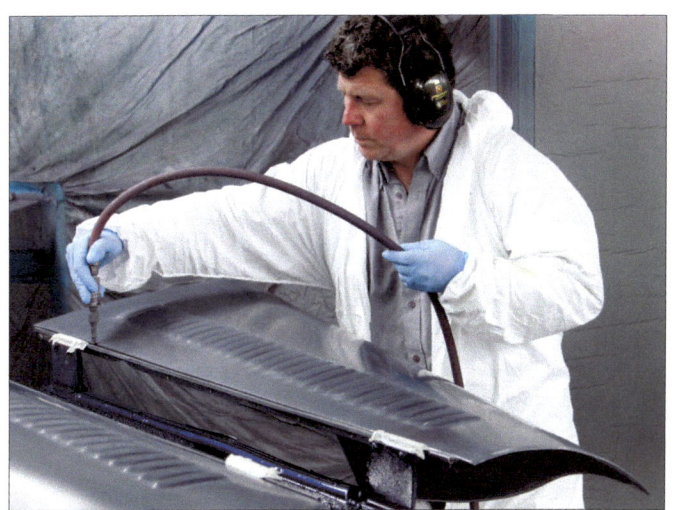

Mark uses compressed air to blow away sanding dust.

The tail lamp plinths are also given a third primer coat.

125

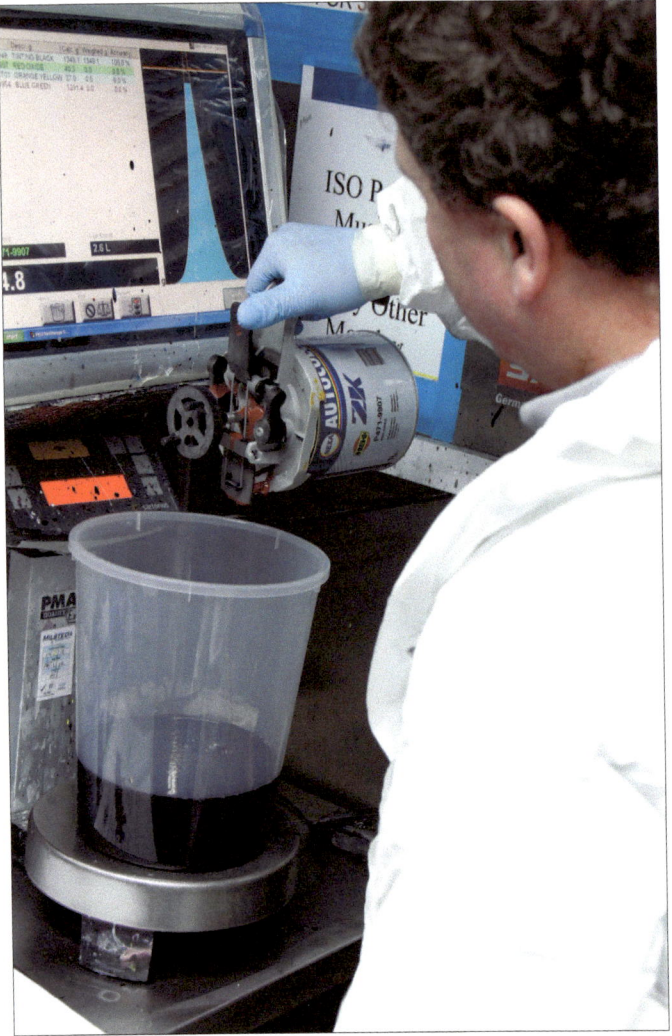

Just be careful not to add a single extra drop …

Paint is mixed to an accuracy of one tenth of a gramme. The right shade of green also calls for shots of orange-yellow and red. The computer assists in metering the ingredients.

That poses no problem; the recipe is in the computer. Just take 1037 grammes of tinting black, plus 31 grammes of red oxide, 20.8 grammes of orange yellow, and 993.4 grammes of blue green. Add 1409.3 grammes of hardener and 797.1 grammes of thinner.

Now, stir well. The result is just over five litres of a historic shade of green, à la Rolls-Royce. In this process, the scales are connected to a computer, which keeps track of the colour components like a rocket launch countdown. The last few grammes of each are exciting, as the last grammes, and of course the very last tenths of a gramme, drop into the container as individual drops, until 0.00 shows on the display.

There follows the first painting of all parts – down to the small housings that will later carry the tail lamps. Then a ten-minute pause, because the first layer of paint needs to 'flash off' – to dry slightly. Soon all parts are resplendent in their exclusive paint colour, but before the drying process is started, a second, thin coat of paint is applied.

Baking the paint at 75°C (167°F) takes 45 minutes. That completes the painting process – and once again, it's almost quitting time. Mark carefully lifts the car from its dolly and puts the other pieces in their appointed places, so that the crew from the trim shop can pick them up. The body stays in the booth overnight to dry the paint; now it really is quitting time. The complete painting process took a total of two and a half days.

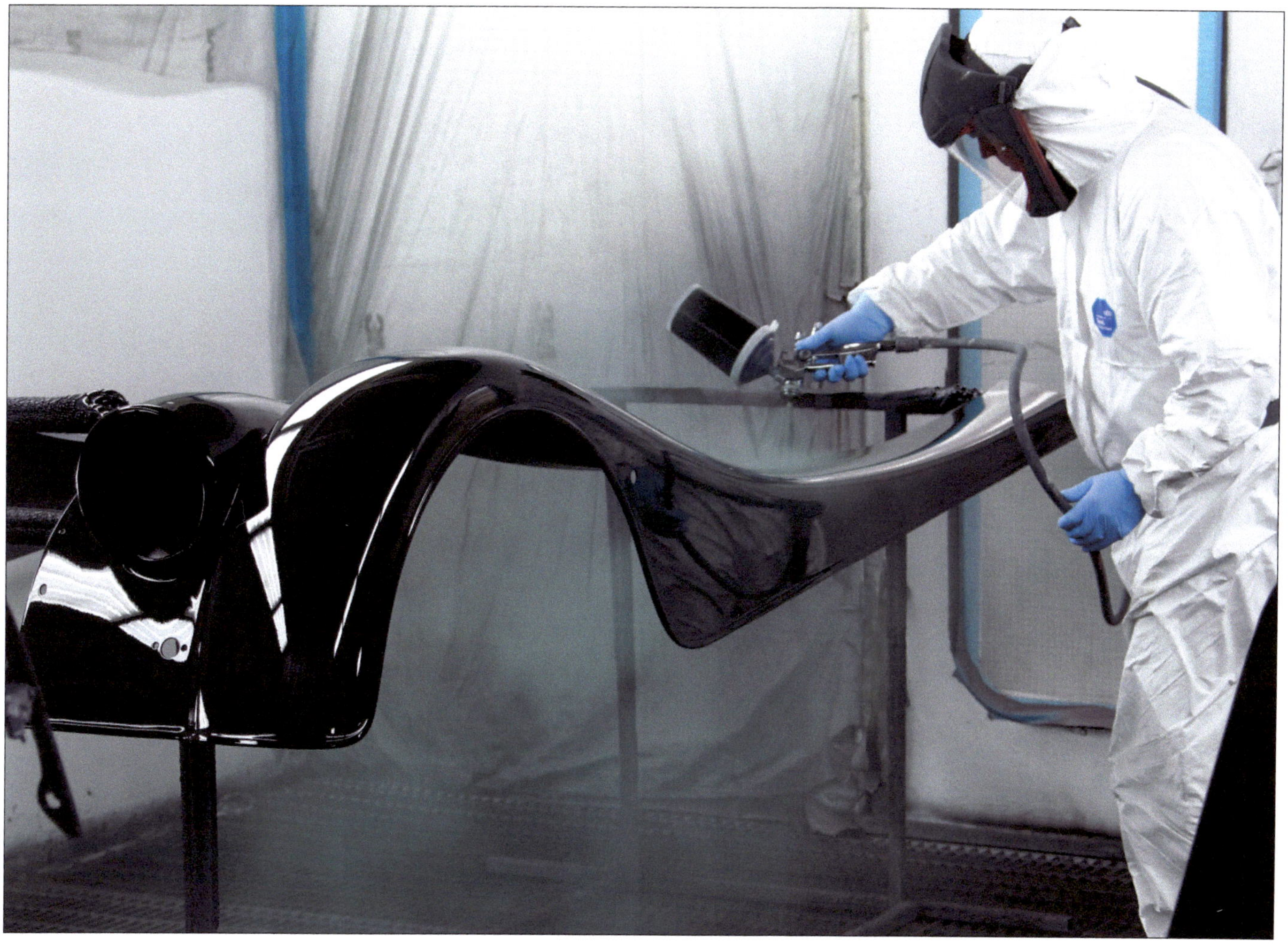

Third day in the paint shop

Finally! Two coats of paint finally give us the desired gloss.

A dazzling outcome.

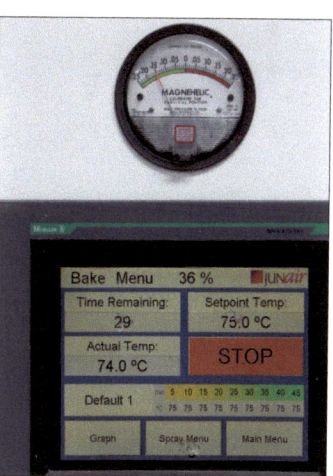

The bake menu calls for a nice warm soak at 74 °C (165 °F). Everything is monitored by a computer.

Day 12

The people from the Trim Shop pick up the car from the Paint Shop.

Trim Shop

When the car is finally pushed into the trim shop at 11:45am, work proceeds apace, at several work stations. Some pieces of carpeting, edged in leather, have already been prepared. Dan, who will concentrate on our car for the next four to five hours, has cut the first pieces of leather for its upholstery and interior trim, and has the foam parts for the seats ready to be covered. He ties all this into a roll and passes this down to other workers, because the seats are built two workstations down in the trim shop. In this, the last installation shop, Dan begins by re-installing the doors.

Right after lunch, Sue – one of the few working women in the Morgan factory – used her sewing machine to hem or add welting to the first cut-out leather and carpet pieces. These are the tops of the doors, the expanding door pockets, the door snubber, and the first carpet pieces for the footwells.

The Trim Shop, 1919.

Meanwhile, Dan has cut more pieces according to patterns, and just after 2 o'clock, he begins to trim the interior. First, the pieces and mating surfaces are given a generous coat of glue.

This is followed by quickly placing the pieces, taking great pains to ensure that there are no wrinkles. Besides glue, the other main fastening method is the air nailer, which hammers away with breathtaking speed.

All elements subject to significant stress are built up in multiple layers. The cover for the rear wheel arches, for example, consists of three layers: a board-like plastic material, foam, and a leather covering.

In this phase, Dan is supported by several colleagues: Taffy cuts the leather panels to cover the transmission tunnel, Paul fits the wooden templates for the inner door panels,

Dan cuts leather to shape, here for the rear wheel arches.

MAKING A MORGAN · DAY 12

Sue sews door pockets.

Dan has to wear a respirator as protection against the adhesive vapours.

This will be the leather covering for the gearbox cover. Taffy cuts out the required shape.

Three layers on the wheel arches provide more padding.

Richard mounts and installs the first of many components such as the radio aerial and assembles the windscreen wiper system from individual parts. He draws parts from a rolling cart provided by the parts warehouse – the four-part trolley.

Meanwhile Dan is well along with trimming the passenger side. For the door sills, he first fits a rigid foam part, as an underlayer. This is topped by more foam, and then leather. The door seal is formed by a windlace – a leather-covered bead incorporating a rubber hose-like component.

Then, leaning far into the footwells, Dan glues the first carpet pieces into place. This completes the passenger side; now he repeats this on the driver's side, but doesn't finish before the bell once again sounds quitting time.

Trim Shop

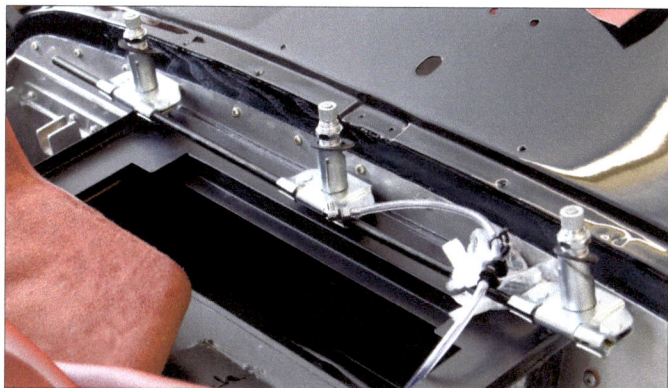

These three spindle posts will carry the windscreen wiper arms. The plastic hose is part of the windscreen washer system.

From muted shades to dazzling colours, everything is on the menu. There is no shortage of leather colour choices.

Leather for the Morgan

On a Morgan, almost everything in the interior that isn't painted or made of metal, is covered in leather. This accounts for much of the car's charm. Several suppliers keep leather, in 150 different colours, on hand. Granted, that's not as many choices as in the case of paint colours, but we can't imagine any other carmaker having a larger selection of leather variations.

Naturally, several different leather colours can be combined, for example in the form of welting or carpet edging in contrasting colours. A traditional two-seat Morgan takes about 20 square metres (215 square feet) of leather. In our case, the chosen colours are MH Red and black.

A gem from the four part trolley: the chrome grille.

Day 13

Richard has his hands full, joining the fuel filler cap to the tank.

Solid! The rectangular steel tube is intended to protect the occupants in the event of a side impact.

Trim Shop II

For reasons unknown, this day starts slowly. Our car is just slightly behind schedule, and the workers in the next section – the trim shop – are already standing by. Richard comes by, from another section of the factory hall, and sets to work. First he connects the fuel hose and tank vent to the fuel filler cap, which requires a great deal of patience and dexterity. Then he turns to the windscreen washer system, installs the nozzles as well as the drive system for the characteristic three wiper arms, whose motor is installed on the driver's side, under the bonnet. To our knowledge, Morgan is the only manufacturer that still uses three wipers.

And what of Dan? He's now occupied with the leather trim on the driver's door, and still has to cover the crash pad in leather and install it. Richard is bolting the side impact bars into the doors.

Next, he mounts a board behind the seats, to serve as a luggage shelf. This now makes easy access to the battery impossible – and with it, outside starting assistance if the car's own battery is too weak. A Morgan driver must never forget to turn off the lights after shutting down the ignition. Long-term storage is also not a good idea. Because there is no other way to start the car, on our own example we elected to install a warning buzzer for the lights.

Taffy has taken another dive into the footwells and is installing carpeting. Richard is installing the sidescreens

Put a lid on it: From this point forward, it won't be quite so easy to reach the battery.

Richard installs the windscreen.

Steph sews the hood.

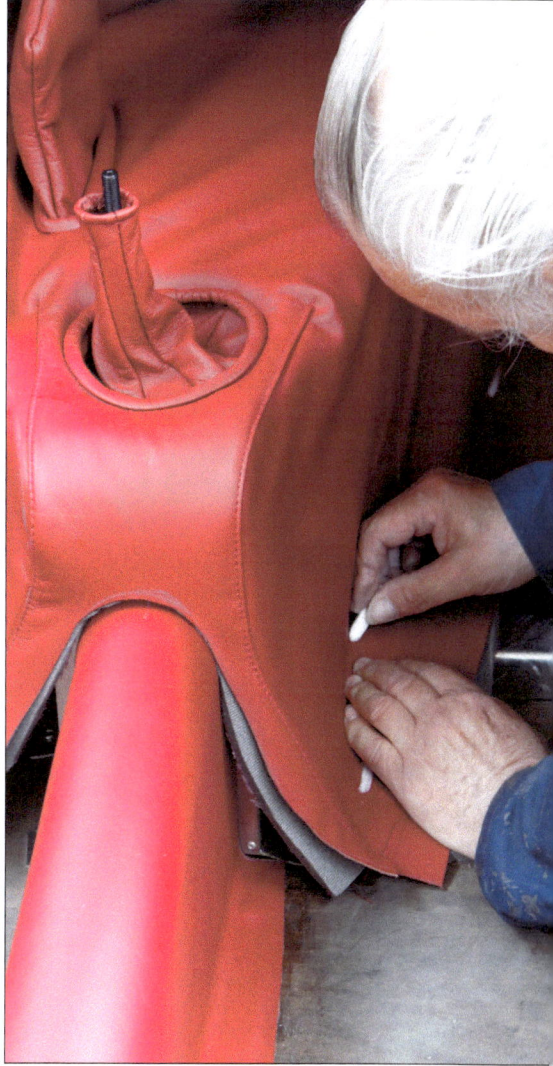

Taffy takes a measurement …

and the tonneau cover. Now we are moving ahead with giant steps. Because two more cars have come out of the paint shop and need to use our space in the first bay, our car is pushed five yards farther ahead, into the second bay. This puts the car that much closer to final assembly. Now Richard installs the speakers; the wiring was already in place.

Dan has withdrawn to his assigned workbench and is covering the last of the parts with dark red leather – the inner door panels, with their door pockets, the covers for the elbows, as well as the rear closing panel of the cockpit, which is normally covered in carpeting. Now Richard is working on the windscreen. He pulls the connecting wires for the windscreen heating system into the interior.

Although this may seem to be an unnecessary extra, violating the purists' ideal of a traditional Morgan, it must be remembered that there is no ventilation directed at the inside of the car's windscreen, and it would very quickly fog up. So, please, no upturned noses at this modern technological achievement.

After the morning break, Richard's work on this particular Morgan comes to a halt. Because Dan is not yet finished with the door panels and small leather covers for the door latch housings, Richard can't continue, so he goes off to help out at another station. We wait for what seems an eternity until, at 11:30am, the eagerly awaited parts are finished and Richard can get back to working on our Morgan.

… and Sue sews the complex transmission cover.

Yvonne sews the seat covers.

Now it's time for the door locks. Installation calls for both the door inner panels as well as the lock covers. Richard checks, several times, that the door latches smoothly, and makes several adjustments. By then, the rear cover, with its openings for the safety belts, is done, and he temporarily lays it in place.

After the lunch break, three women workers turn their attention to 'our' car. Steph sews the soft top (or 'hood,') which takes a good hour. Yvonne makes the cover for the transmission tunnel, while Sue sews leather bindings around various pieces of carpeting.

I have to jump from one work station to another so as not to miss a thing. Work also continues on the car itself; Taffy applies leather to the propshaft cover.

A half hour later, after Yvonne has finished the complex leather part, she goes to work on the seat covers – a task that will take a good two hours. Meanwhile, Taffy test-fits his leather part and marks those areas that require rework.

Next, Sue, after selecting a different thread colour, takes over the final stitching of the multi-layer piece. In between, she checks in on Yvonne and Steph. The pleated centre section of the seats is already recognisable, and the soft top, of fine mohair, is nearing completion.

After Taffy adds some fastening parts, he completes the installation of the gearbox cover. Now, even without seats, the interior looks quite liveable – generously trimmed in leather and with most carpeting in place. Meanwhile, Steph has finished the soft top.

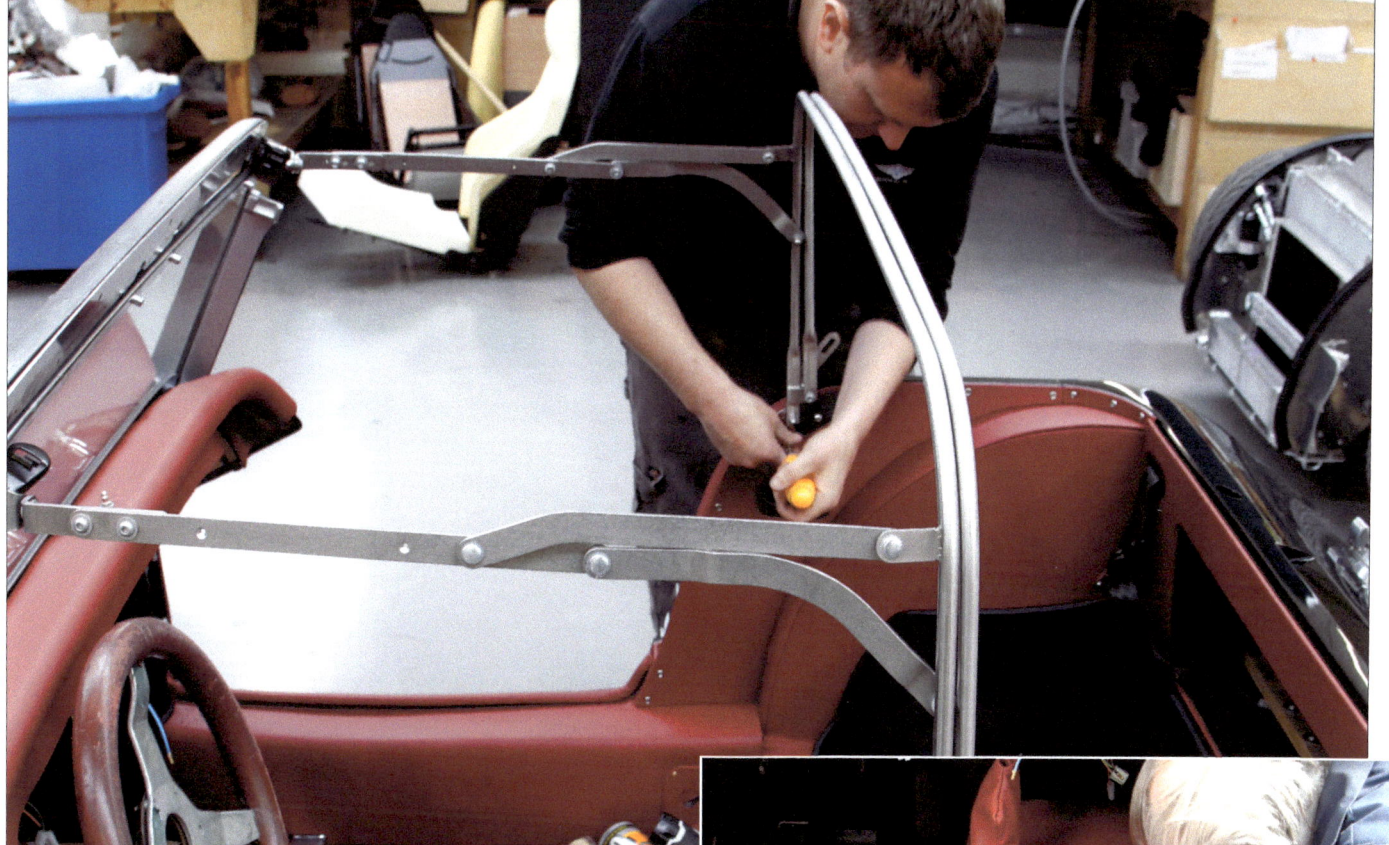

The hood frame is supplied by the Machine Shop.

Leather and carpeting transform the raw structure into a refined, comfortable driving environment.

We have reached another highlight. The car is pushed to the last section of the trim shop, before going to final finish. There, the seats will be fitted, complete with the finished covers, and the soft top completed and fitted. Steven already has the soft top mechanism in hand, having obtained it from the machine shop, and proceeds to install it. A board is taped to the windscreen; I won't comprehend its purpose until later.

Then Steven takes the rearmost hood bow with the rear latch mechanism and covers it with black felt. This keeps him occupied until quitting time. Tomorrow, the car is supposed to be completed to the point where it can be test driven. At least, that's what Steve, the department head, has promised us. The weather forecast looks good. We'll see if they can make the deadline.

Day 14

First test fit of the soft top.

The soft top

Once again, orders are to be present at 7am. This morning, Ben has taken over the work on the custom-fitted soft top. He bolts the rear portion of the frame to the rest of the mechanism, and does the first tests of the soft top. The soft top material is tacked to the frame, and then, with tiny nails, attached to the board that had been taped to the windscreen and whose purpose had been a mystery to me until now.

Then he pulls the fabric taut, and again installs temporary tacks and nails. The entire operation is reminiscent of custom tailoring. Instead of a pincushion on his arm, Ben holds nails in his lips – and chalk in his hand to mark fastening points for the hood material.

The front hood bow does not quite fit the seam, which is somewhat too far forward. Some adjustments are made. Having served its purpose, Ben removes the board from the windscreen, and again opens the hood. Gradually, the trim shop is filling up. Some workers arrive early, to have a chat with colleagues and talk about the previous day's play in the rugby league.

The soft top

Ben Jones –
the racer

Ben is 36 years old. He was born in South Wales. At the age of 11, he and his parents moved to Malvern. After completing school, he did not complete an apprenticeship, but instead worked in a design studio. After that, he joined Steve Simmons, a former Morgan employee, who had established his own business, customising Morgans as well as other luxury marques, and soon made a name for himself in the scene.

When a spot opened up at Morgan, Ben seized the opportunity, as factory wages were considerably higher. That was twelve years ago. Now, among other things, he makes Morgan hoods. He loves the tradition and the old craftsmanship, and feels completely comfortable at Morgan. Ben has been living with his girlfriend for several years. In his spare time, he loves fast machinery – he rides a motorcycle and races karts.

The board taped to the windscreen serves to anchor the taut soft top material.

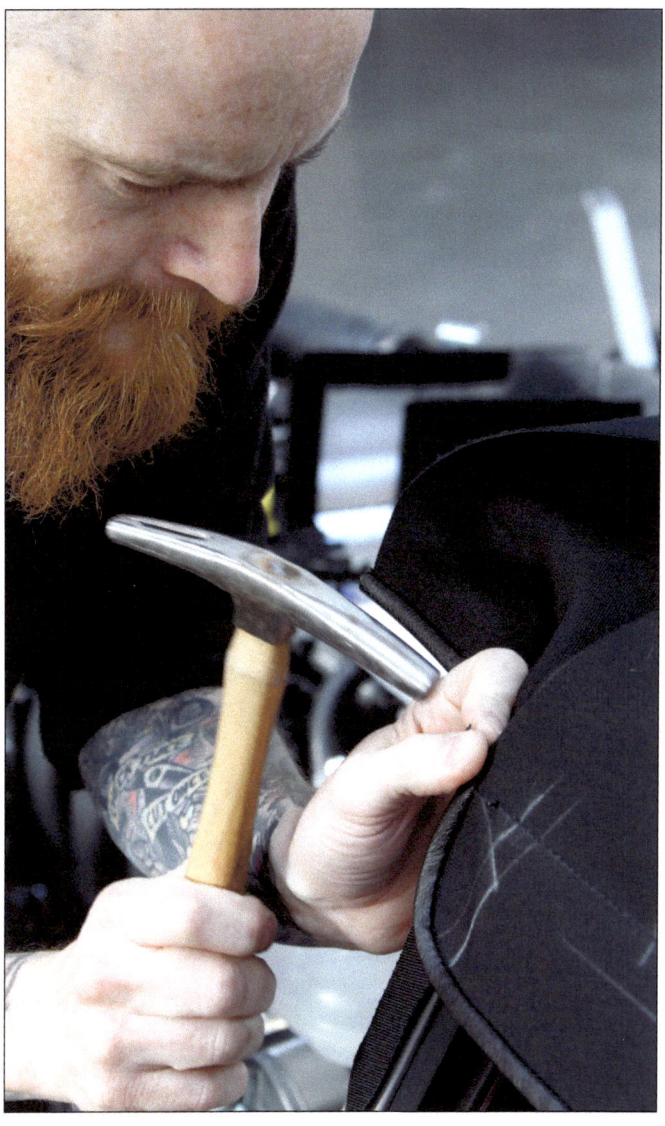

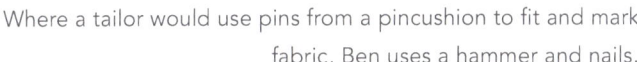

Where a tailor would use pins from a pincushion to fit and mark fabric, Ben uses a hammer and nails.

Ben now fits the tonneau cover, which Sue has sewn. Hood, tonneau cover and the lower parts of the sidescreens are usually all of the same colour; customers can choose from eight PVC (vinyl) and seven mohair shades. The ladies at the sewing machines, joined by Ben as the only male representative, have over time gravitated to specialised tasks: Steph makes hoods, Yvonne does mostly seats, and Sue makes tonneau covers.

Ben test fits the tonneau cover, so that Sue can finish hemming its edges.

Neatly masked and well coated with glue.

Tonneau cover

A tonneau cover is an element typical of British roadsters, used to cover the car's cockpit when the hood is not raised, or not present. It begins at the bottom edge of the windscreen. On the driver's side, it can be opened by means of a zipper, so that the rest of the car can remain covered while driving. Closed, it serves mainly as a protection against dust, leaves, rain, or prying eyes. It's not one hundred percent water-tight; water can enter through the zipper, for example. It is, however, quite able to withstand a drizzle.

Like the hood, the tonneau cover is test-fitted. Ben uses chalk to mark all the areas that he still needs to trim away, and the location of fasteners. After the final trimming, done with scissors while fitting, he turns the hood over to Sue, who hems all the edges with her sewing machine.

The hood is again removed – it's only held on by four screws, after all – and finally fastened to the frame. Ben begins at the back; both sides have been coated with contact adhesive. The temporary air-nailed fastening is now replaced by a clean line, the old clamps are removed. Ben trims the excess fabric at the new clamping line. He does the same at the front of the hood, which when raised is attached to the windscreen frame. Then he neatly applies crepe tape to the rear of the hood, to mask it against excess adhesive, which he applies generously. After the first layer of glue has dried, Ben applies a second layer. He does the same to the rubber seal, which will be attached at this point.

Seals are also glued to the front of the hood frame. These are especially important, as they prevent entry of water at the windscreen frame. Wherever the contact adhesive doesn't quite do the job, Ben applies industrial-strength Super Glue (cyanoacrylate adhesive). From experience, it's known that after frequent use, parts of the seal will loosen, which makes

Tonneau cover

The soft top fits perfectly.

Craig covers the lower sections of the sidescreens with mohair.

closing the hood much more difficult. It's a good idea for the on-board tool kit to include a small bottle of Super Glue, and an appropriate piece of wood, to push the seal back into place.

The tonneau cover is only missing the fasteners that attach it to the car. This job is soon completed. Now comes the big moment, the final test: how does the hood fit? Is it too tight to close smoothly? Are there any creases? After the hood is screwed into place, we see that it fits perfectly.

Now we're only missing the supplied sidescreens, with their traditional sliding windows. They need some minor adjustment, and their lower halves covered with fabric or vinyl. Craig has taken on this task. First, the shapes of both screens are matched to the doors on hand. Because each door is unique, the screens, sourced from an outside supplier, can't fit perfectly – a situation that Craig corrects with a pair of tin snips.

Then he uses chalk to trace the screen outline on a double layer of covering material, and cuts out the shapes, generously allowing some material for seams. Steph sews the doubled layers together, following the chalk line. Craig uses a knife to cleanly cut the excess material along the seam, and turns the resulting shell inside out, forming the pocket for the lower portion of the sidescreen. Craig applies some adhesive to join the two parts, and … finished. A matching storage bag – naturally adorned with an embroidered Morgan crest – is obtained from the well-supplied warehouse.

Day 14, and it's nearly lunchtime. We had secretly hoped that the test drive would take place today, before quitting time. But

Step by step, the indicators, gauges, and operating controls are connected. The interior mirror is lying atop the dashboard, and tiny wiper arms hang from the windscreen, waiting for installation.

Meanwhile, Jon continues fitting the electricals.

Jon installs the electronic control unit (ECU). Now the fusebox has a much tidier look.

Will fits seat covers.

our subject still looks like it's a long way from being a finished car. It still hasn't had its seats fitted, and some parts still need to be installed in final finish, the last station in the trim shop. And today, the sun is finally shining – perfect conditions for a little drive.

Electrics and small parts

The good news is that a bay and workers are now available in final finish. Jon comes over and begins to complete the electrics, while Will starts on the seats. Like the sidescreens, the seats' metal frames and moulded foam parts are outsourced. Will just has to assemble the parts and cover them with the sewn leather covers. In all, a job that takes two hours – for both seats.

Electrics and small parts

Jon has set aside the steering column stalks, instrument panel, and windscreen wipers for installation.

Jon connects the instrument panel to the wiring harness.

Jon installs both oxygen sensors on the engine exhaust and connects them to the electronic control unit (ECU), a nondescript little box installed below the junction and fusebox. The ECU is the electronic brain of the engine. Finally, there is some semblance of order in the fusebox. Fuses are installed, the remaining wires attached, and the fusebox has a clean, orderly appearance. Later, a decal giving the function of the 50 fuses is attached to the fusebox cover.

Now – finally – the car is pushed around the corner to final finish, where all the remaining parts are installed and it turns into a complete car. It is, in effect, nearing the finish line – but we can't see the chequered flag yet, and the clock is ticking relentlessly. We'll write off our hopes of doing the test drive today; meanwhile, the weather forecast for tomorrow is less than encouraging.

From his box of wonders, Jon removes another small box, whose installation location won't be revealed here. This is the anti-theft system, which prevents the engine from being started without the ignition key – 'hot-wired,' to use the vernacular. The installation kit also includes the inside rear view mirror – a functional, if somewhat questionably styled, dimming mirror from the Ford parts bins – and the windscreen wiper arms and multifunction steering column stalks, also sourced from Ford.

Finally, Jon connects the starter motor and programs the keys for the anti-theft system. A quick twist of the ignition key ... the starter works and the battery is supplying current. The car is jacked up even as Jon is still mounting the tiny windscreen wiper arms.

After the afternoon break – we still have two and a half hours to quitting time – Jon is assisted by Peter, Lyndon, Steven and Alex. The entire crew is working on 'our' car.

Jon pours 18 litres of fuel from a canister into the tank. That should be enough for the test drive. After extended cranking, the first starting attempt is successful. It's drivable, at least.

Earlier, Jon had fitted the varnished wooden dashboard with instruments, warning lamps and switches. Compared to our own personal 2011 model, the location of individual instruments has changed. Two smaller gauges, for water temperature and fuel tank level, are now directly in front of the driver.

An oil pressure gauge, like the one in our personal car, is absent. We also find it unfortunate that the traditional round

Lyndon installs the tail lamps.

instruments of the venerable old British firm of Smiths are no longer being installed. They don't always work reliably (our oil pressure gauge measures everything imaginable, with the exception of actual oil pressure) but visually, they are a better match to the cockpit than the VDO instruments presently being installed.

While Will was working on the second seat, a number of other things happened to the car simultaneously: The instrument panel was installed, the heating controls were connected to the heater, the radiator and fan were installed, the rear wings were attached along with tail lamps, door sill trim and safety belts.

And again the soft top is unscrewed to make it easier to work at the back of the car. Meanwhile, farther forward, Jon installs the radio on top of the gearbox cover. This makes it anything but easy to operate, but it also makes it less visually intrusive, something that's apparently important to most Morgan drivers who order this option. If only it weren't for that traitorous aerial ...

Work at the rear is made more difficult by the absence of an opening for the spare wheel. This mostly affects Peter, who not only installs the third brake lamp, but also a bin for temporary storage of the sidescreens. The existence of such a bin is new to us, and we've also noted it on other Morgans being built. It is lined with carpeting and so deep that even my long arms won't reach all the way to the bottom. This bin significantly expands the storage space behind the seats.

Steven fills the cooling system and pressurises it by means of a pump to check for leaks. Next, he mounts the first wing. At

Electrics and small parts

Will brings the other seat.

The cooling system is filled and pressure-checked for leaks.

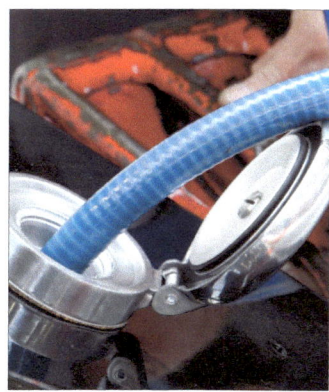

18 litres (c.5 gallons) of fuel are poured into the tank. That should be enough for the test drive and delivery.

the moment, there are two workmen lying underneath the Morgan.

Meanwhile, Peter attaches the seats. Quitting time is rapidly approaching, and the way it looks right now, a test drive with the finished car is out of the question. Peter and I run up some overtime, so that the Morgan can at least be finished tomorrow.

Lyndon has installed the headlamps in the front wings, and Peter's task now is to bolt on the second wing, route and connect the wiring for headlamps and turn signals. As both wings are now in their appointed places, Peter can attach the front number plate mounting bracket, the bumper overriders, and the cowl, which will house the radiator grille.

In the past, no overriders were used on the narrow body, as space is at a premium on its narrow front. More recently, this option has been made possible – and our front aspect is now complete. Peter turns his attention to the three-part exhaust system, again spending considerable time beneath the car. Apparently, he has spent the entire day down there.

Finally, he mounts all four wheels again. But where are those beautiful eared knock-ons? They simply can't be found, and there's nobody in the warehouse, so Peter puts on ordinary nuts to allow the car to be taken off its jackstands. Finally, it's quitting time for us, too. Now the car actually looks like it could be taken on its test drive tomorrow. Let's hope that the weather co-operates.

Space under the car is tight.

Naturally, the exhaust system is only accessible from below.

Finishing touches

Day 15

For the final touches, the workers need to dive deep into the car.

Finishing touches

The sky is overcast, but at least it's dry. Light rain is forecast for the afternoon. Our leading actor still needs a few minor touches in makeup; the radiator grille is installed, the bonnet latches are adjusted to allow easy opening and locking, the soft top is installed again and closed. FINISHED! At exactly 9:30am, the Morgan is ready for its test drive – fitted with a temporary steering wheel and a protective mask for the front end, so that there is not even the tiniest stone chip in the immaculate paintwork when the car is finally delivered.

The soft top is back in place, and the car is almost finished.

It's alive! A new Morgan embarks on its test drive, and test driver Jon lets it roar.

Test drive

Test driver Jon has five years of experience at his job. The wonderfully scenic standard test route, over lonely backroads, extends for 22 miles (35 km). Normally, Jon completes the circuit in about 40 minutes. It will take a little longer today, because we want to take a few pretty pictures of the route.

We follow Jon in our 'media Morgan,' enjoying the idyllic landscape of the British Midlands around the Malvern Hills. The test drive goes without a hitch; no rattles in the car, all controls operate correctly, good road manners, well-behaved steering and faultless brakes. Jon is most satisfied, and writes up his test report. Back at the factory, he adjusts the headlamps and installs their chrome trim rings. Next, he checks the front wheel camber using an electronic gauge. He has to make a few adjustments until both wheels read 0.0.

Now he can run the car up on the lift and tighten the tie rods, without crawling under the car. Done! Simultaneously, Keith and John thoroughly inspect the car, from top to bottom – among other things, looking for leaks and fasteners that aren't quite tight. But everything is faultless.

Test drive

A brief stop for photos.

Taking last pictures.

The test pilot is satisfied and brings the Plus 4 back into the factory …

... where the headlamps are adjusted ...

... as is front wheel camber ...

... where the workmen are happy to be able, for once, to stand up while working beneath the car.

The new Plus 4 is allowed off the lift, and Jon installs the production steering wheel – again a Ford product with a visually somewhat questionable design, but one that is mandated for safety reasons. After tea time, Jon again takes the car out for a brief run to check the steering after the final tightening of the tie rods.

The Morgan is now ... well, not quite finished, because Steve still has to do a visual check and – finally – the technical approval.

Steve conducts the final quality check and signs the technical approval.

Day 16

Undersealing is done out-of-house, by STP Motor Services.

Underseal

That was a long Whitsun holiday, for us as well as the Morgan. On the sixteenth day, for starters, nothing is happening. The Morgan is parked, rather forlornly, alongside the building where the pre-delivery inspection will be carried out, waiting for the next step.

The next stage in its story is a company called STP Motor Services, whose workers apply the undersealing to all Morgans. Beginning in the mid-1980s, undersealing was available as an option, and has been standard equipment from the mid-90s. In the past, STP operated directly within the Morgan works, in the space that is now occupied by the design center. Now the firm is located in a small industrial park directly behind the Morgan works – literally a stone's throw, or five minutes by car.

STP is a bottleneck in the production flow, because the firm can only handle one car in the morning and one in the afternoon. 'Our' Morgan is scheduled for the afternoon, and we won't see it again until the following day.

Underseal

Covered with a 'bra' and equipped with temporary registration the new Morgan roars through the neighbourhood.

First, the car is carefully masked off so that the viscous black wax won't touch any of the painted surfaces or metal components. Subsequent removal of any excess would require a major effort. To even be able to apply the wax, it has to be heated. Even after cooling, it doesn't harden completely, but rather stays flexible. Its consistency is such that even after damage, such as a stone impact, it undergoes a sort of self-healing process, and small cuts close up by themselves.

Day 17

One final, thorough inspection of the paintwork.

PDI

The calendar tells us it is Tuesday, May 26. A historic date, because what will probably become the most famous Morgan of all time (after Le Mans winner TOK 258) will finally be finished today, ready for delivery. This is two working days later than originally planned, but still well ahead of our planned departure from Malvern Link.

Today, there will be a final inspection, and once again the air is heavy with suspense. A flaw may be detected at the last moment, requiring extensive rework. That means we would have to extend our stay at the Foley Arms Hotel – which would be workable, but of course unplanned. Not by Morgan, nor by us.

A brief detour to quarantine. The reason is marked: a trim problem.

The pre-delivery inspection takes place in the next building, and normally takes four to five hours. The department contains a number of brightly lit individual bays. First, the paintwork is thoroughly cleaned. With a daylight-bright hand lamp, the smallest defects are sought out. Even 'our' Morgan shows an occasional fault, marked with a special pen. Cars that are not completely faultless are placed in a quarantine bay.

The car is also sent to quarantine if any flaws are detected – for example, during the test drive – that cannot be remedied immediately or that fall under the responsibility of a department other than the final finish shop. A whiteboard indicates which departments have to step up to carry out rework operations on the quarantined car.

Fortunately, for us, there are only a few paint flaws that can be removed on the spot. That is done in the afternoon, right in the PDI bay. For hours, the car is cleaned, buffed, polished ... until truly all flaws have been removed and not even a single fingerprint mars the shimmering paintwork.

Early in the afternoon – at precisely 2:30pm – the car stands ready for transfer to the delivery warehouse in Kentvale. But that won't happen today, and so we bid adieu to 'our' Morgan, and all of the workers whom we have

A final machine polish …

… followed by one more pass with the polishing cloth, and 'our' Morgan is ready for delivery.

Also tiny paint flaws need to be addressed.

annoyed for the past three and a half weeks, and whom we have taken into our hearts.

Fully confident that the car will truly be finished today, we have already checked out of our hotel and stowed the luggage for our month-long English sojourn in our own personal Morgan. Wistfully, we drive through the works gates. A last wave … and off we go, under sparkling sunny skies, in an open car, to friends in Yorkshire, where we will relax before taking the ferry from Hull to Rotterdam.

We wish the buyer of 'our' Morgan every joy with the car, for decades to come, and of course safe, accident-free motoring. We eagerly await the new owner's reaction to 'his' or 'her' book.

The Morgan factory in Malvern Link, 2015.

EPILOGUE

We've put three and a half weeks behind us – an interesting time, an intense experience. Never before has a team of authors peered over the shoulders of the Morgan crew, for so long, or so intensely – up close, with a camera, from the first bolt on the chassis to the final polishing.

It should be high time for change at Morgan – after more than a century, in which nothing much has changed. However, reassuringly, as ever, at the centre of the Morgan philosophy are the workers, many of whom draw on decades of experience – and an extraordinary level of craftsmanship and skill, as can hardly be found anywhere else. It comes as no surprise that at Morgan, training for some specialties takes as much as five years. Regardless of their task or work station, the capabilities of these men and women have earned our deepest respect.

In those seventeen days, a small work of art was created, a masterpiece such is not to be found anywhere else on this Earth – at least, not in the automotive world. In keeping with the principle of maintaining tradition while cautiously embracing modernity, Morgan creates a vehicle for enthusiasts, at once quirky and odd, yet which has maintained its identity virtually unchanged throughout the past sixty years of its production. And will continue to maintain that, as long as laws and regulations in individual export markets and in its mother country England allow. Actually, the traditional Morgan should be on the World Cultural Heritage list – has anyone bothered to nominate it yet?

Hopefully, it will never come to pass that Morgan will turn the helm over to some nameless investors who would exchange more than a century of manufacturing heritage in Pickersleigh Road, Malvern Link, for a shiny new factory, or replace the artistry practised by traditional craftsmen for the soulless operation of modern machines. Or even that the shape, with its incomparable wooden frame, is altered.

Naturally, there must be detail improvements, nobody would object to those. Perhaps the sidescreens could be free of rattles against the windscreen frame, or if the roof were halfway water-tight ... But please spare us – we are speaking of the traditional Morgan – from the scourge of even more plastic in the interior, on-board computers, or other electronic foolery that isn't absolutely necessary for the act of driving.

Leave us with the pure, unadulterated joy of driving a classic – even if it is factory new and therefore considerably easier to maintain than a decades-old car. Anything else, we can buy elsewhere, 'off the peg.' This pure driving experience can only be had in a Morgan, or a true vintage car – and the latter is seldom capable of everyday service.

At Morgan, cars are built by people, not machines. We had the privilege to be allowed to make their acquaintance, to accompany them throughout their shift – even if it began at seven in the morning. To look over their shoulders, but also to join them in their favourite pub for a beer. We have experienced how they put their passion and dedication into their work – even when they are just about to enter retirement.

Epilogue

We saw their pride in being privileged to work for Morgan – a company that is like a huge family to them. Nobody ever complained that they had to crawl under the car or into a footwell to tighten a simple bolt, or that a part was too heavy or that the breaks were too short.

We were immersed in a relaxed, happy atmosphere. We heard jokes, songs, conversations – as work allowed. We saw supervisors who didn't continuously keep an eye on the workers' hands, but rather got their own hands dirty as they pitched in. Everyone lent a hand, when more hands were needed.

And from the beginning, we did not feel like intruders, but rather were welcomed by the workforce. We found a home within the old red brick walls, were allowed to breathe in the atmosphere, to experience the spirit that makes Morgan so special. For us that was an honour.

We hope that our story has given you, the reader, a glimpse of that spirit, that in reading, you have sensed the uniqueness of the company, and that if you aren't already a proud owner of a Morgan, at least have an urge to experience this feeling for yourself. Caution is indicated; Morgans are most definitely addictive!

Credits

Brian Iles Collection: 7
Dennis Rushton Collection: 19
Dorothy Morgan album: 8
DPPI: 29
Dr. Jake Aldersen collection: 18
Gordon Acock: 17
George Morgan albums: 8, 9, 11, 21
Martin Webb: 6, 9, 13, 15, 20
Morgan family album: 8, 11
Morgan family archives: 12, 23
Morgan Motor Co.: 11, 12, 13, 14, 15, 16, 17, 19, 20, 21, 22, 23, 24, 26, 27, 31, 34, 36, 37, 39, 40, 41, 51, 54, 129
Reproduced with kind permission of the Ordnance Survey: 18
Steven's Directory/Malvern Library: 6
Stilltime.net : 19
The Autocycle: 22
The Cyclecar, December 4,1912: 21
Worcester County Records Office: 17
Motor Cycle: 18

ACKNOWLEDGEMENTS

It took us several hours to bid farewell to all the workers we had gotten to know, and who gave us so many very personal farewell gifts to take on our journey. It is difficult for us to single out individuals for special thanks.

We'll try anyway. Our special thanks go to Steve Morris, the managing director of Morgan, who needed no more than five minutes at the 2014 Geneva Auto Show to agree to this project. Our German publisher, Franz-Christoph Heel of Heel Verlag, who was also quick to climb aboard and join us in making this project a reality. And, naturally, for their dedication to the project, down to visiting the factory near the end of the production as well as a pleasant farewell gathering in our favourite pub. To our editor, Jost Neßhöver, for his creative assistance and keeping us on a long leash.

James Gilbert, who organised the works visit and gave us every possible support. Naturally Martyn Webb, who, among other duties, is assembling the Morgan archives, is the author of a highly detailed history of Morgan, and granted our every wish with respect to historical photographic material. Without him, we would probably have been lost in the archives. We thank him for his hospitality – in a pub, at any of several weekend hillclimbs, or at a private dinner in his home. In that regard, also warmest regards to his wife and his friend Graham. Many thanks also to Dave Edwards of Superform Aluminium for an interesting plant tour and a stimulating discussion.

And thanks to all the workers, the men and women of Morgan, who suffered our peering lenses and prying questions and our general getting in the way. They showed superhuman patience and composure: John, Nick, Stuart, Brett, Sam, Jamie, Nigel, Karl, Vince, Nic, Steve, David, Mark, Jon, Sean, Graham, Richard, Conrad, Edward, Martin, Jason, Kevin, Blake, Marcin, Philip, Mick, Chris, Adam, Wayne, Ross, Jim, Dan, Sue, Taffy, Paul, Steph, Yvonne, Steven, Ben, Craig, Will, Peter, Lyndon, Alex and Keith. If anyone is wondering why there are so few names, well, there are several Adams, Marks, and Stevens. They know who they are.

"**Beautifully produced ...** a complete in-depth look at the new Morgan 3 Wheeler" - *Total Kit Car*

"If you are a Morganist, you **will want to own a copy**" - *Octane*

"**... a must-have book ...**" - *MOG Magazine*

THE MORGAN 3 WHEELER
BACK TO THE FUTURE!

A century after the car's first appearance, Morgan returned to its origins with the 3 Wheeler. This book examines how it was created, how it became Morgan's best-seller, why it is (or isn't) your kind of vehicle, what it's like to drive, and its strengths and weaknesses. Also looks at factory improvements since the 2011 launch, and other modifications.

Hardback • 25x25cm • 144 pages • 101 colour and b&w pictures
ISBN: 978-1-845847-63-0

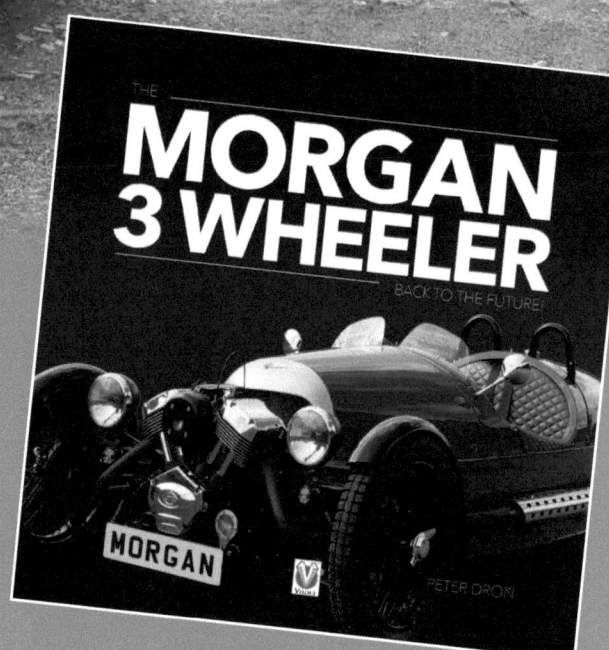

For more info on Veloce titles, visit our website at www.veloce.co.uk • email: info@veloce.co.uk • Tel: +44(0)1305 260068